EDMOND HALLEY

THE MANY DISCOVERIES OF THE MOST CURIOUS ASTRONOMER ROYAL

David K. Love

Essex, Connecticut

An imprint of Globe Pequot, the trade division of
The Rowman & Littlefield Publishing Group, Inc.
4501 Forbes Boulevard, Suite 200, Lanham, Maryland 20706
www.rowman.com

Distributed by NATIONAL BOOK NETWORK

British Library Cataloguing in Publication Information Available

Library of Congress Cataloging-in-Publication Data Available

ISBN 9781633888913 (cloth : alk. paper) | ISBN 9781633888920 (epub)

∞™ The paper used in this publication meets the minimum requirements of American National Standard for Information Sciences—Permanence of Paper for Printed Library Materials, ANSI/NISO Z39.48-1992.

Before Edmond Halley, comets were dreadful things, full of dire import. They foretold doom and destruction, the coming of plague, the death of a ruler, the fall of a kingdom. But the Astronomer Royal changed all that . . . and *much* more.

To Charlotte, Stella, and Oscar,
in the hope that you will be able to see
Halley's Comet in 2061

© The Royal Society. A portrait of Edmond Halley, aged about 33, by Thomas Murray, painted in about 1690 and on display at The Royal Society in London. The plaque beneath the portrait states that he was clerk of The Royal Society from 1686 to 1698 and secretary from 1713 to 1721. *Reproduced by kind permission of The Royal Society*

CONTENTS

PREFACE

The story of astronomy is endlessly fascinating. Thanks to our growing astronomical knowledge and in the space of only about 500 years, we have moved from a position of believing that the Earth was at the very center of a relatively small Universe to one in which we now know that our planet is no more than the tiniest of specks in an unimaginably massive (and perhaps infinite) Universe.

Edmond Halley was an important part of this story. He is rightly famous for his prediction that the comet of 1682 that he had observed would return about every 76 years. In doing this, he gave the clearest possible demonstration of the validity of Newtonian physics. His other colossal achievement was to bring about the publication of Isaac Newton's *Principia* (which embodied the very physics that enabled his prediction), something that Newton on his own would never have managed. But his accomplishments went well beyond these two things, as this book aims to show.

Halley was somebody who fitted a huge amount into his long life, and his work on his many undertakings often overlapped. To give the reader a clearer idea of what he achieved in each field, some material (in particular chapter 3, on Halley's Comet) is dealt with in a single section, even where this means covering later dates than those that appear in subsequent sections. For those who would like a strictly chronological list of dates, this is provided in appendix B. The book as a whole is nevertheless as chronological as I could make it, subject to this constraint.

The danger of a biography, which I hope I have avoided, is that it can descend into a hagiography. This is a particular difficulty with Halley because he genuinely was widely (but not universally) liked; however, it certainly doesn't help that we know hardly anything about his personal life. Nor does it help that the biographies written within a few years of his death do not have a single unkind word to say about him. Probably the worst things that can be said of him are that he may possibly have plagiarized the work of others (see chapter 5), that some of his theories and mathematical calculations were erroneous, and that on occasion he was a little too willing to do Isaac Newton's bidding.

A book that recorded every detail of what Halley did would be noticeably longer than this one, so I have restricted it to those of his accomplishments that I deem to have been particularly important or interesting.

I would like to thank many people for their assistance while I was writing this book. I am especially grateful to Nick Lewis, who read through each chapter as I wrote it and made numerous helpful comments. Nick was a particularly appropriate person for this task, as he graduated from the same university as Halley (Oxford) and in much the same subject. My daughters, Rachel Love Carney and Catherine Love Soper, and my wife, Dr. Elizabeth Emerson, also read through early drafts and between them put forward several useful ideas.

A huge thank-you must go to Rupert Baker, the librarian at the Royal Society in London. He not only gave prompt and helpful replies to all my e-mails but also pointed me in the direction of several primary source documents in the possession of the Royal Society. Ellen Embleton, also at the Royal Society, was supportive in digging out a number of illustrations for the book. I am also grateful to Royal Astronomical Society librarian Dr. Sian Prosser and History of Astronomy Society librarian James Dawson for their help.

I am deeply grateful to Professor Richard Ellis, who willingly read through a draft of this book and offered up a great deal of very helpful advice. My heartfelt thanks also go to Professor Robert Walker of Oxford University, who gave me a conducted tour of Halley's haunts in Oxford. I owe a huge debt of gratitude to the late Dr. George Wilkins and his successor, Dr. Keith Orrell, for running the History of Science group at the Norman Lockyer Observatory during a span of many years. I have learned a lot from its numerous meetings.

A special word of thanks must go to Sandra Greaves, who kindly dug out some well-sourced information on the wonderfully named Admiral Sir Cloudesley Shovell, the possibility of whose grisly death had been predicted by Halley some years previously—see chapter 7. Thanks too to Ginny Dawe-Woodings, St. Paul's school archivist, who supplied me with what little information the school had on Halley's time as a pupil there.

Finally, I am particularly grateful to Jake Bonar at Rowman & Littlefield for spotting that a popular science Halley biography was well overdue and asking me to write it. I am also grateful for the help of Nicole Carty and the other members of the Rowman & Littlefield team.

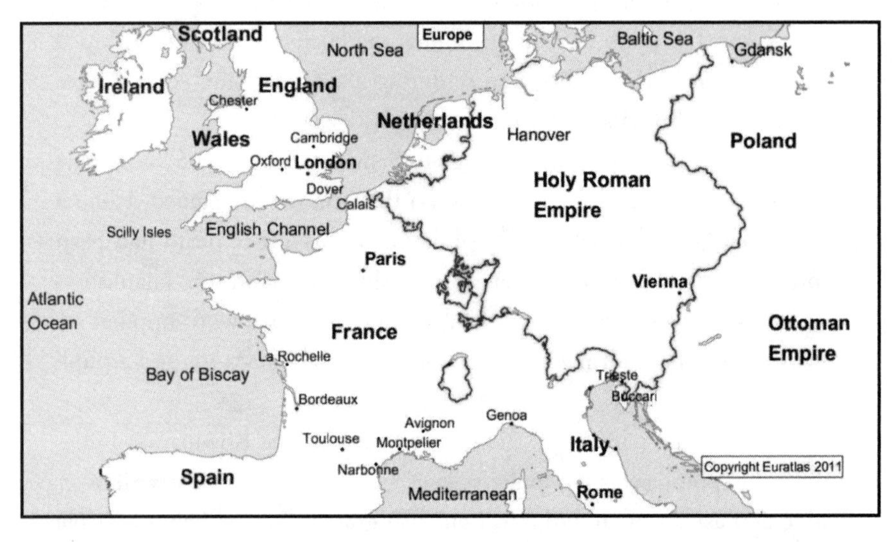

Map of Europe as it was in 1700, showing places visited by Halley. ***Reproduced by kind permission of Euratlas Maps***

INTRODUCTION

Astronomy before Halley

Astronomy is the oldest of the physical sciences. It has a long and complex history that is not easy to summarize in only a few short paragraphs. The best place to start is with the ancient Greeks, who seem to have made the first serious attempts to give a physical explanation (albeit incorrect) of what exactly was happening in the night sky.

To almost all the ancient Greeks who considered the matter, it was obvious that the Earth was spherical in shape (not flat) and was both stationary and (more or less) at the center of everything. It was surrounded by an immense and very distant sphere, the celestial sphere, to which all the fixed stars were attached. This sphere rotated around the stationary Earth about every 24 hours, which explained why the stars drifted across the sky as the night went by.

But as well as these fixed stars, there were seven wandering bodies in the sky that moved independently of the celestial sphere. This was obviously the case because they were constantly changing their positions relative to the background stars on the celestial sphere. These seven wanderers were the Sun, the Moon, Mercury, Venus, Mars, Jupiter, and Saturn. (Our modern word "planet" is derived from the ancient Greek word for "wanderer.") They were clearly in a different category from the fixed stars.

So ideally, the Greeks wanted to be able to explain the Universe in the way proposed by Plato (ca. 428–348 BCE), in which the Sun, the Moon, and the planets all moved around the Earth in perfect circles at uniform speed, as shown in figure I.1. Unfortunately, the reality refused to conform to the Platonic ideal. The Sun and the Moon more or less appeared to do so. However, the central problem was the fact that the planets didn't appear to move around the Earth in the same direction all the time. Most of the time, they moved from west to east (relative to the background of the fixed stars), but from time to time, they appeared to move backward for a few weeks or months before resuming their original courses.

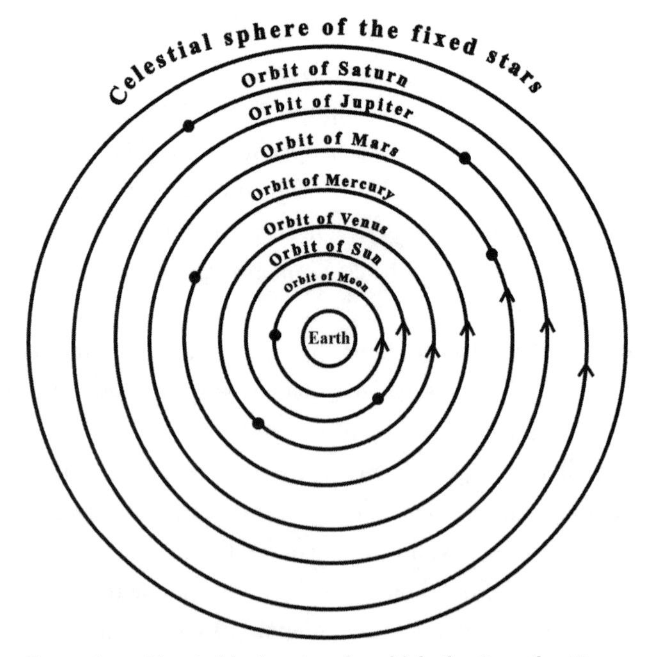

Figure I.1. Plato's ideal system in which the Sun, the Moon, and the planets all move around the Earth. ***Created by author***

The ancient Greeks put forward various complicated explanations for this strange behavior, almost all predicated on the idea that the Earth was stationary and (more or less) at the center of the Universe. The idea of one ancient Greek, Aristarchus of Samos, who lived in the third century BCE, that the Earth orbited around the Sun, was not taken seriously. Earth-centered mathematical models culminated in the work of the ancient Greek astronomer Ptolemy (ca. 100–170 CE).

Ancient Greek astronomy came to an end with Ptolemy, and the baton was picked up by the Arabs and the Persians, who both preserved and extended the ancient Greek inheritance. The brilliant poet and mathematician Omar Khayyam (1048–1131), to give just one example, measured the length of the year to a remarkably high degree of accuracy. Many of the star names that we still use today (Aldebaran, Algol, Altair, Betelgeuse, and so on) are Arabic in origin. (The prefix "al" is simply the Arabic definite article.) And if it had not been for the Arabs, numerous valuable Greek manuscripts would have been lost.

But the big breakthrough did not come until Nicolaus Copernicus (1473–1543), who was born in Toruń in what is now part of Poland. He found a simple explanation for the strange movements of the planets. He realized that if the Earth was moving around the Sun, just like the other planets, then the reason why the planets sometimes appeared to move backward was simply because that corresponded to the period of time when they were being overtaken by the Earth in its orbit. Much the same illusion happens when a fast train overtakes a slower train going in the same direction. To someone on the fast train, the slow train seems to be moving backward.

Unfortunately, although his idea was conceptually far simpler than the complicated ideas of the ancient Greeks, his continued attempts to rely on uniform circular motion meant that his predictions of future planetary movements were little better, if at all, than those of the Ptolemaic system.

A further breakthrough came with the German astronomer Johannes Kepler (1571–1630). His painstaking use of the very accurate planetary observations of the Danish astronomer Tycho Brahe (1546–1601) led him to his three laws of planetary motion. These included the key discovery that all the planets, including the Earth, moved around the Sun in ellipses rather than circles. They enabled Kepler to predict future planetary positions to a previously unheard-of degree of accuracy and provided a compelling argument for the view that the Earth went around the Sun rather than the reverse.

Tycho Brahe also constructed an atlas of about 1,000 stellar positions, which Halley was later to make use of; however, Tycho's measurements were made without the benefit of the telescope, which was not invented until a few years after his death. So in Halley's time, there would be considerable scope to improve on Tycho's measurements.

Kepler lived in the same period as Italian physicist Galileo Galilei (1564–1642), who took full advantage of the newly invented telescope. His discovery of the phases of Venus demonstrated conclusively that Venus orbited around the Sun rather than the Earth. His discovery of the four large moons of Jupiter (along with Simon Marius, a German astronomer who independently made the discovery at about the same time) clearly showed that not everything revolved around the Earth. Galileo was not the first person to view the night sky through a telescope, but he made a number of significant discoveries with it and was usually able to see their significance more easily than others.

So the world of Edmond Halley, which we are about to enter, was one where most educated people in Europe had now come to accept that the Earth rotated on its axis once every 24 hours and revolved around the Sun once every year. Increasingly, our Sun was correctly seen as just another star among many rather than being at the center of the Universe, as Copernicus and Kepler had imagined.

But were there any underlying and unifying laws of nature that lay behind the apparently random laws that Kepler had discovered? And could the use of experiment and observation in a wide range of other areas provide us with a better understanding of how the world operated?

Step forward Edmond Halley.

1

HALLEY'S EARLY LIFE AND CAREER

Edmond Halley* was born on October 29, 1656,† in the tiny village of Haggerston. At that time, the village consisted of little more than a country house belonging to his wealthy father.[1] Haggerston lay within the parish of St. Leonard's church, Shoreditch—now a small part of the London Borough of Hackney. (The bells of St. Leonard's were one of the six sets of church bells immortalized in the famous English nursery rhyme "Oranges and Lemons."‡)

* Halley's name was normally spelled this way and is normally pronounced as in "galley" or "valley." Other spellings were used during his lifetime. His first name is almost always spelled "Edmond," notwithstanding the spelling of "Edmund" in the portrait of him, shown at the front of the book. It seems to have been a common feature of names in that period that people were untroubled by different spellings.

† It has sometimes been suggested that Halley was born in 1657 rather than 1656, but the grounds for this suggestion are not particularly strong. All dates falling in Halley's lifetime are based on the Julian calendar—see appendix F for an explanation of this. Mainland European sources tend to quote Halley's birth date as November 8, 1656, reflecting their use of the Gregorian calendar that England was very slow to adopt.

‡ For those who don't know the words to "Oranges and Lemons," which refers to several central London churches:

> Oranges and lemons, say the bells of St. Clement's.
> You owe me five farthings, say the bells of St. Martin's.
> When will you pay me? Say the bells of Old Bailey.
> When I grow rich, say the bells of Shoreditch [i.e., St. Leonard's]
> When will that be? Say the bells of Stepney.
> I'm sure I don't know, says the great bell at Bow.

Halley's parents had married in the upmarket church of St. Margaret's, Westminster, next to the Abbey, early in September,[2] less than two months before he was born. However, it seems likely that they had also married in a civil ceremony several months—or even years—beforehand, as the legal conducting of marriages at about that time had temporarily been taken away from the clergy.

Halley was the first of three children, but his sister Katherine[3] seems to have died in infancy, and his brother Humphrey died in 1684 or possibly earlier. His father, also called Edmond, was a well-connected soap maker in Winchester Street* in central London who also owned significant amounts of property elsewhere in London. Of his mother Anne,† née Robinson, we are certain of nothing beyond her name.

At the time of Halley's birth, Isaac Newton was a youth of 13, just beginning his education at Grantham Grammar School in Lincolnshire. Newton's future rival, Robert Hooke, was a young man of 21. Johannes Kepler had been dead for the previous 26 years and Galileo Galilei for 14 years.

This was a tumultuous period in English history. The country had just emerged from a bitter civil war, lasting from 1642 to 1651, between Parliamentarians ("Roundheads") and Royalists ("Cavaliers").‡ King Charles I had been beheaded by the Roundheads in 1649, after which the army general Oliver Cromwell became the Lord Protector of what turned out to be only a temporary republic. The monarchy was restored under King Charles II from 1660, a couple of years after Cromwell died.

The young Edmond Halley was both bright and ambitious. At some unknown date,[4] his father sent him to the prestigious St. Paul's School, founded more than a century earlier by the then dean of St. Paul's

* Now "Great" Winchester Street, in the City of London (the original part of London, founded by the Romans, and now London's financial center) off Old Broad Street.

† Usually spelled "Anne" in older papers but sometimes spelled "Ann."

‡ The Roundheads were famously described as "right but repulsive" and the Cavaliers as "wrong but wromantic" in the Sellar and Yeatman book *1066 and All That*.

Cathedral in London on a site next to the cathedral. The school was closed down in 1665 because of the Great Plague of London, which killed an estimated 100,000 people, very roughly one-quarter of the total population of the city. Then, while the school was still empty, it—along with the cathedral—burned down in the Great Fire of London of 1666, when Halley was nearly 10. The Great Fire devastated a large part of the center of London. More than 13,000 houses and 80 churches were destroyed. Halley's father's house on Winchester Street was dangerously close to the area destroyed by the fire, only a few streets away, but had a lucky escape. Some of his other properties were not so fortunate.

After a brief spell when the headmaster (Samuel Cromleholme) moved to premises in Wandsworth in south London, probably taking many pupils with him, the school returned in 1671 to a new building on the St. Paul's site, where Halley would have received much of his early education.[5] Here, he acquired his language skills (Latin, Greek, and Hebrew[6]) as well as a sound foundation in geometry and astronomy. The curriculum at St. Paul's seems to have been a particularly broad one because he also obtained some knowledge of the science of navigation,[7] something that was to stand him in good stead in later years. At some stage, probably while at school, he also picked up a working knowledge of French, Spanish, and Italian.[8] In a sign of things to come, he also became captain of the school at the age of 15. His mother Anne died late in 1672 while he was still at St. Paul's, and his father later remarried. This second marriage, to a woman named Joane, was later judged by many to have been imprudent because of her unfortunate ability to spend her husband's money.

OXFORD UNIVERSITY

In the summer of 1673, when he was 16, Halley moved on to Queen's College at Oxford University. At Queen's, although he was interested in all the sciences, he found he was confirmed in his earlier decision to pursue astronomy. Professional astronomers in the modern world tend to be either theoreticians or observers. Throughout his life, Halley effortlessly managed to be both. He was barely 19 when he wrote an original paper on how to compute some of the key elements of a planetary orbit. While at Oxford, he also observed an occultation of the planet Mars by the Moon (when the Moon passes in front of Mars as seen from the Earth). Both his paper and his observations were published in the *Philosophical Transactions* of the prestigious Royal Society.[9]

Halley conducted observations of the planets Jupiter and Saturn, using equipment he had brought with him to Queen's and paid for by his doting father, and showed that they deviated from the positions predicted by the best available tables of the time, a problem he became determined to rectify. He was confident enough of his results that, in a letter of March 10, 1675, he felt able to write to the newly appointed (and first) Astronomer Royal,* John Flamsteed (1646–1719), about these errors.[10]

THE ROYAL GREENWICH OBSERVATORY AND THE LONGITUDE PROBLEM

Flamsteed was shortly to be placed in charge of the Royal Greenwich Observatory (in what is now part of southeast London) that was to be built later in 1675. This observatory might never have been built had it not been for the fact that King Charles II had a French mistress, Louise de Kérouaille. In 1674, Louise used her position to introduce Charles to the French astronomer Sieur de St. Pierre,[11] who may have been another

* Strictly speaking, Flamsteed was appointed as "Our Astronomical Observator." The formal title of Astronomer Royal did not come about until much later. Nevertheless, Flamsteed is by common consent regarded as the first Astronomer Royal.

Figure 1.1. The Royal Greenwich Observatory as it is today; it now functions as a museum. ***Photo courtesy of the author***

of her clients and who, more importantly, had claimed to be able to solve the vexed problem of determining longitude at sea.*

The problem for ships at sea and out of sight of land, which Sieur de St. Pierre had claimed to be able to solve, was that they had no direct way of determining their longitude—in other words, their east-west position. Latitude, their north-south position, was relatively easy to calculate, assuming good weather. This could be determined by observing the angle above the horizon of either the Sun at midday or the Pole Star at night. (By no more than a fortunate coincidence, the Pole Star happens to lie almost directly above the North Pole, hence its name. So measuring its altitude in degrees above the horizon gives a very good indication of

* It is interesting to note that, through their only son, Charles II and Louise were the distant ancestors both of the current King Charles III's first wife, Diana, and of his presumed mistress and second wife, Camilla. Diana and Camilla were therefore distantly related to each other. More background on Louise de Kérouaille is given at http://www.unofficialroyalty.com/louise-de-kerouaille-duchess-of-portsmouth-mistress-of-king-charles-ii-of-england.

latitude.) But nobody had yet come up with a satisfactory way of determining longitude, even though this was becoming increasingly important for navigational reasons. Ships out of sight of land knew their positions fairly accurately in the north-south direction (latitude) but could only estimate their positions in the east-west direction (longitude) through the rather inaccurate dead reckoning method.*

St. Pierre proposed a method that depended on measuring the position of the Moon relative to the background stars—this position would be different at different longitudes, which could then be calculated.[12] Charles II set up a committee (that included Robert Hooke and Sir Christopher Wren, both of whom will feature later in this story) to examine this proposal. The committee rejected it as unworkable for the simple reason that it was not at that time possible to predict the position of the Moon against the stellar background with sufficient accuracy. Instead, they recommended to Charles II that an observatory be set up and a suitable observer (Flamsteed) appointed to measure stellar and lunar positions more accurately, with the aim of trying to solve this longitude problem. Christopher Wren suggested the site at Greenwich and also acted as architect for the building. As a result of befriending Flamsteed, Halley was one of the members of the party that inspected the site in June 1675. Flamsteed laid the foundation stone on August 10, 1675. His brief was

> the rectifying of the tables of the heavens, and the places of the fixed stars, in order to find out the so much desired longitude at sea, for the perfection of the art of navigation.[13]

In 1676, the Royal Greenwich Observatory was completed, and Flamsteed began work. Although he was a very capable astronomer, he was also a pedantic, narrowly religious, solitary, and rather vindictive

* Dead reckoning is an unsatisfactory means of determining a ship's current position from its known starting position together with its estimated speed and direction of travel from this starting position.

man, and his initially warm relationship with Halley was to degenerate in later years. Flamsteed's measurements of stellar positions were already being duplicated by the famous Polish astronomer Johannes Hevelius (1611–1687) in Danzig (Gdansk), but his results would act as a useful check on both men's measurements.

What they were attempting, to measure stellar positions more accurately, is illustrated in figure 1.2. For this purpose, it is useful to imagine that the stars really are fixed to a gigantic sphere centered on the Earth and rotating around the Earth about once every 24 hours. By projecting the Earth's latitude and longitude coordinates onto this sphere, we get a celestial frame of reference against which the positions of the stars (in the celestial equivalent of latitude and longitude) can be measured. Angular

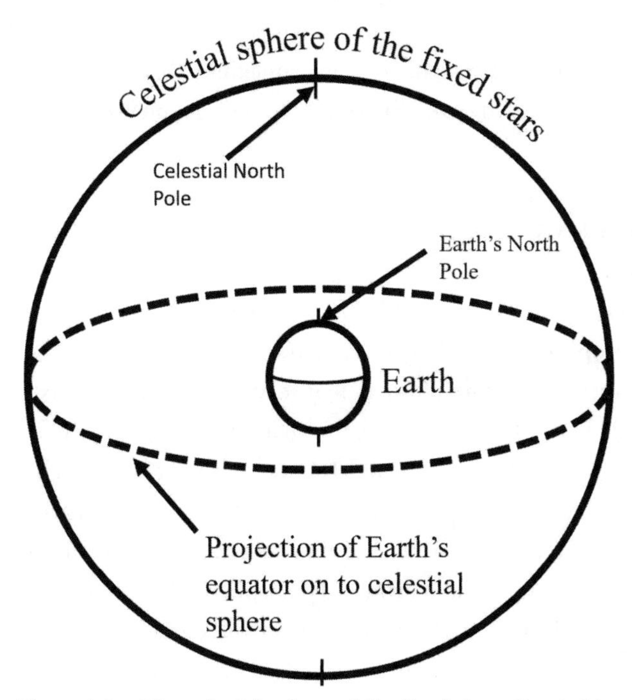

Figure 1.2. The celestial sphere of the fixed stars. ***Created by author***

distances between stars are measured in degrees and fractions of a degree. The Sun, the Moon, and the planets are all constantly moving against this background of fixed stars.

It was while he was still at Oxford that Halley spotted how he could make a valuable contribution to astronomy as well as making a name for himself. Flamsteed and Hevelius were observing and cataloging only the stars that could be seen from the Earth's Northern Hemisphere (and specifically from about 51° and 54° north of the equator, respectively). There was no point in repeating their work. As Halley said, if he had done so, he "would only be cackling stupidly among such stately swans." However, Flamsteed and Hevelius were unable to observe stars that were visible only from the Southern Hemisphere. All Halley had to do was to find a suitable observing spot south of the equator, get himself and his equipment there, and make accurate observations of the positions of a few hundred of the brightest Southern Hemisphere stars, something that had not been done before, and he would have achieved something genuinely useful.

ST. HELENA

After looking at a few possible sites, Halley decided on the small and remote tropical island of St. Helena in the South Atlantic. The island's principal claim to fame today is that it was the location of Napoleon's final exile from 1815 to 1821, more than 100 years after Halley's visit. Critically, St. Helena was located 16° south of the equator. This was far enough south to be able to observe all the stars in the Southern Hemisphere, but it would also enable him to measure their positions relative to some stars in the Northern Hemisphere. The island was owned by the English, who had placed it under the control of the East India Company,[14] which used it as a stopping-off point on the route between

England and India. (The shortcut provided by the Suez Canal would not be available for another 200 years.)

Thanks to his time at Oxford, Halley had contacts in high places and was now able to make use of them. He passed on to Sir Joseph Williamson (1633–1701) and Sir Jonas Moore (1617–1679), both of whom held important positions in the English government, his plans to visit St. Helena. They in turn approvingly mentioned the matter to King Charles II. For his part, Charles then instructed the East India Company to provide Halley, his equipment,* and a friend, James Clarke,† with free transport

Figure 1.3. Map showing the location of St. Helena in the South Atlantic. ***Wikimedia Commons***

* "His equipment consisted of a sextant of 5½ feet radius with telescopic sights, a 2 feet quadrant, a pendulum clock, two micrometers and several telescopes" (MacPike, 1937, 39).

† Various spellings of this surname are used in the source documents.

to St. Helena on their ship the *Unity*. To help with general finances, Halley's father provided him with a grant of £300 a year, a princely sum for those days. (For comparison, as Astronomer Royal, Flamsteed only ever received £100 a year from the government, with no allowance for any astronomical equipment—although he did also get his accommodation at the observatory thrown in free.)

So Halley left Oxford University before obtaining his degree and set sail for St. Helena in November 1676, arriving there in February 1677. Three months seems to have been a fairly normal time span for getting there. The time was not wasted, as it was on this voyage that he learned more of the seafaring and navigational skills that he would make ample use of later in life. It was only on his arrival that he was to discover that St. Helena was not the ideal observing spot that he had hoped for and that he had been led to expect. The problem was the bane of all observational astronomers—the cloudy weather. This meant that he was not able to obtain the positions of anything like the number of Southern Hemisphere stars that he had anticipated. Annoyingly, the cloud made its appearance over the island but not over the surrounding sea.

In a letter of November 1677 to his patron, Sir Jonas Moore, he complained that

> now, when I expected to be returning, I have not finished above half my intended work; and almost despair to accomplish what you ought to expect from me. I will yet try two or three months more, and if it continue in the same constitution, I shall then, I hope be excusable if in that time I cannot make an end.[15]

This was not his only problem on the island. The governor, one C. Gregory Field, probably not realizing the importance of what Halley was attempting, was deeply unhelpful. This was in spite of having been instructed to provide Halley and his friend with accommodations while on the island as well as giving them all necessary assistance. Perhaps this imposition simply irritated him. Following several complaints about the "ill-living" of the governor, Field was removed from office at about the same time as Halley was leaving St. Helena.[16]

Notwithstanding these problems, Halley managed to plot accurate positions for 341 stars during a period of about a year, and he published these in his *Catalogus Stellarum Australium* in 1679, as soon as possible after his return home. He plotted these stars relative to more northerly stars that had been observed by the Danish astronomer Tycho Brahe and that could also be seen this far south. He even referred to his own catalog as a "supplement" to Tycho's.

To curry royal favor, Halley added a new constellation to the southern skies, consisting of 12 stars: "Robur Carolinum," or Charles's Oak. (Constellations are no more than random groupings of stars that are thought to form a memorable pattern.) This was in honor of the occasion in 1651 when the then future king Charles II hid in an oak tree to escape the Roundhead troops led by Oliver Cromwell after Charles's heavy defeat at the battle of Worcester. The constellation has not survived—today, its stars are incorporated into the modern constellations of Carina and Vela.

Halley was neither the first nor the last astronomer to try to gain royal approval by naming a celestial object after an appropriate ruler. Nearly 70 years earlier, in 1610, Galileo had used his telescope to discover the four large moons of Jupiter. At the time, he lived in Padua, near Venice, but wanted to move to Florence to work for the Medici family, the rulers of Florence. So he named the moons the "Medicean planets." He got the job, but the name didn't survive—today, the moons are simply named Io, Europa, Ganymede, and Callisto after characters in Greek legends.

Much the same trick was attempted by William Herschel in 1781 when he discovered the planet Uranus, which he wanted to name "Georgium sidus" after King George III. George was suitably impressed and granted Herschel a modest pension, but again the name has not survived.

TRANSIT OF MERCURY

Halley didn't only catalog stellar positions. In October 1677, he also observed a transit of the planet Mercury as it passed across the face of the Sun. Transits of the inner planets Mercury and Venus come about because all the planets in the Solar System orbit around the Sun in much the same plane. In other words, you can draw the planets, their orbits, and the Sun on a flat piece of paper without too much inaccuracy. There is no need to attempt a three-dimensional model. However, if Mercury and the Earth orbited in *exactly* the same plane, then transits of Mercury would occur regularly about every 116 days, each time Mercury overtook the Earth in its orbit around the Sun. But more often than not, Mercury will pass either slightly above or below the Sun as seen from the Earth, so transits happen only every few years. (The very first successful prediction of a transit of Mercury, for November 1631, was made by Kepler in his Rudolphine Tables and was then observed by the French astronomer Pierre Gassendi.)

Halley probably already realized that such transits (whether of Mercury or Venus—and preferably Venus, which is both closer to the Earth and bigger) provided a way of accurately measuring the Earth–Sun distance for the very first time. He would later make a detailed proposal as to how this should be done, to be discussed in chapter 7. In Halley's day, only relative distances within the Solar System were known with any certainty. So, for example, it was known that the average distance between Mercury and the Sun was 0.39 times the average Earth–Sun distance, but that was all. What we don't know is whether this transit method was an original thought on Halley's part, which is perfectly possible, or whether, perhaps more likely, he had already read of the idea in the writings of James Gregory, a brilliant Scottish mathematician who had died in 1675 at the tragically young age of 36. If the latter, it is a valid criticism of Halley that he never acknowledged any debt to James Gregory. He was, however, good friends with his nephew, David Gregory, another talented mathematician and astronomer, which implies there was no bad feeling in the Gregory family over Halley's perhaps unacknowledged proposals.

OMEGA CENTAURI

Halley also observed the globular cluster now known as Omega Centauri. Realistically, you need to be as far as possible south of a latitude of about 35°N to see this. As he was at a latitude of 16°S, he was very

Figure 1.4. Omega Centauri. ***European Southern Observatory (ESO)***

well positioned. Although it had been observed before (it is mentioned in Ptolemy's great astronomical work, the *Almagest*), he was the first to recognize it as an object that was not a single star. We now know that globular clusters are roughly spherical clusters of densely packed stars, numbering anything between ten thousand and several million. There are roughly 200 of these, and they form a halo around our Milky Way Galaxy. Omega Centauri is the largest of these that we know of and contains approximately 10 million stars. The largest cluster visible in the northern sky contains only a few hundred thousand stars. Coincidentally, this was also discovered by Halley many years later in 1714 and is now known as Messier 13, the Great Cluster in the constellation of Hercules.

HALLEY'S PENDULUM CLOCK

Another discovery that Halley made while on St. Helena was that the length of the pendulum in his clock needed to be shortened, compared with what it had been in London, to give the correct time.[17] This was not the first time that this effect had been observed in a clock that had been moved closer to the equator—the French astronomer Jean Richer had seen the same thing in 1672 while in French Guiana to observe the planet Mars.

We now know that this correction is necessary because gravity becomes very slightly weaker as you move toward the equator. There are two reasons for this. First, the Earth is not a perfect sphere; it bulges at the equator and is flattened at the poles, so an object at the equator is farther away from the center of the Earth and will therefore feel a smaller gravitational pull. Second, the Earth spins on its axis once every 24 hours, so a point anywhere on the equator travels farther and has to move faster than a point at, say, London. This will mean that the outward centrifugal force at the equator is greater, so also slightly reducing the weight of an object. The combined effect is tiny, but if you want to lose weight by a tiny amount, you should move to the equator. But do note that the total amount of matter in your body won't change at all—just the gravitational pull on your body.

MAGELLANIC CLOUDS

Halley was also able to see the Magellanic Clouds, two fuzzy white patches in the southern night sky that can be seen only from the Southern Hemisphere. We now know that these are two satellite galaxies in orbit around our own Milky Way Galaxy. We shall come across them again in chapter 9.

RETURN TO ENGLAND

Halley finally left St. Helena in March 1678 and arrived back in London in May. King Charles was so impressed with his efforts (and, no doubt, with the constellation created in his honor) that he told Oxford University to award him an MA degree without the need for further study. This was possibly the first time that a first degree had been awarded for significant research work. Halley was also elected as a Fellow of the Royal Society in November at the very young age of 22, having been proposed by Sir Jonas Moore, who had been instrumental in getting him to St. Helena.

One rather amusing report of his that reached the Royal Society shortly after Halley's return was that the air on the island "was extremely temperate and helpful." By way of example, he related the case of a 55-year-old husband and his wife of 52 who had moved to the island from England, and the wife—apparently as a result of these temperate and helpful conditions!—had become pregnant while he was staying there.[18]

It was probably a garbled version of this story that led to the rumors of Halley's sexual improprieties while on the island.

THE ROYAL SOCIETY

The Royal Society had been established in 1660 by scientific luminaries such as Sir Christopher Wren, Robert Boyle, John Wilkins, and Robert Hooke. It is one of the world's oldest surviving national scientific societies.[19] (The German National Academy of Sciences was founded in 1652, and the French Académie des Sciences was founded in 1666.) At the time, the term "natural philosopher" was often used for people taking a scientific approach to problems—the word "scientist" wasn't coined until as late as 1833. But the scientific aims of the Society were clearly summed up in its motto, "Nullius in verba," which is usually translated as "Take nobody's word for it." This was a clear indication that the Society's approach was to use observation and experiment to determine truth rather than to accept what some authority figure had said in the past. Halley was to play a prominent role in the Society in the years to come.

Figure 1.5. Edmond Halley's signature, recorded in The Royal Society's Charter Book, in which all Fellows sign their names. © *The Royal Society*

JOHANNES HEVELIUS

Halley's first undertaking on the Royal Society's behalf was to journey to the port of Danzig (Gdansk), on the mouth of the River Vistula, to visit the eminent and now elderly Polish astronomer Johannes Hevelius, who had been cataloging Northern Hemisphere stars. Hevelius was so highly regarded that he had been elected as a Fellow of the Royal Society in 1665. Unfortunately, Flamsteed's cataloged positions didn't always agree with those of Hevelius. Flamsteed and Hooke blamed this on the fact that the rather old-fashioned Hevelius refused to use newfangled telescopic sights on his instruments, preferring instead to rely on the naked eye. In this, it has to be said that Hevelius was completely wrong; telescopic sights undoubtedly improve the accuracy of observations. Halley wrote to Hevelius on November 11, 1678, saying that he was sending him a copy of his Southern Hemisphere catalog and that he was hoping to call on him shortly.

The visit, of about eight weeks, took place between May and July 1679. Halley took his own measuring devices, complete with telescopic sights, and the two of them began their observations together on the evening of Halley's arrival. When they carried out measurements together, each using their own instruments, Halley was surprised by what he found. Hevelius was able to obtain results from his equipment that were better than Halley had expected was possible for an observer who was using only the naked eye. As he explained to Flamsteed,

> But as to the distances measured by the [sextant], I assure you I was surprised to see so near an agreement in them, and had I not seen, I could scarce have credited the relation of any; verily I have seen the same distance repeated several times without any fallacy agree to 10" [a very tiny angle equal to 1/360 of a degree].[20]

One can only conclude that Hevelius must have had exceptionally good instruments and exceptionally acute eyesight. Even so, telescopic sights would have improved the reliability of his results.

Halley had the ability to get on with almost everyone he met, and Hevelius was no exception. He described Halley as "a most welcome guest, a very upright man, and very devoted to the truth."[21] For his part, Halley seems at the time to have had a genuine respect for his host. However, some years later, after making a more wide-ranging study of Hevelius's observations, Halley changed his mind and referred to him in rather disparaging terms, calling into question the accuracy of his observations.*

A tragedy struck Hevelius only a couple of months after Halley's departure. His observatory, his library, and several surrounding buildings were burned down in a huge fire. This must have been a heavy blow, one from which he never fully recovered. As soon as he heard the news, Halley penned a letter to Danzig expressing the sincere hope that Hevelius himself had survived the fire.

It was in this letter that we learn that Halley had promised to send Hevelius a silk dress for his wife, Elisabetha, from London.[22] Hevelius's wife, his second, was some 36 years younger than him and is seen as one of the first female astronomers because of the substantial amount of assistance she gave her husband. By all accounts, Elisabetha was also a very attractive woman. Some decades later, a rumor emerged that her relationship with Halley during his visit had become more than merely platonic.[23] It is certainly not impossible that she, then aged only 32, found herself physically attracted to this eager young man (perhaps more so than to a husband who was by then in his late sixties and who suffered from arthri-

* "As to Mr Hevelius we heare as yet no farther from him, and I am very unwilling to let my indignation loose upon him, but will unless I see some publick notice taken elsewhere, let it sleep till after his death if I chance to outlive him, for I would not hasten his departure by exposing him *and his observations* as I could do and truly as I think he deserves I should" (MacPike 1937, 96, emphasis added).

tis) and that Halley willingly reciprocated. We shall never know. But not many men would send an expensive silk dress to somebody else's wife.

Hevelius managed to rebuild the observatory, acquire new albeit inferior instruments within a couple of years, and continue with his work. But he died a few years later in 1687 at the age of 76, leaving behind Elisabetha, by then still only 40 years of age, and three daughters.

FRANCE AND ITALY

In December 1680, at the age of 24, Halley took time off for a tour of France and Italy together with an old schoolfriend, Robert Nelson. His father generously continued to supply his £300 annual allowance.[24] The channel crossing to mainland Europe didn't go at all well. As he said in a letter to Robert Hooke,

> I got hither the 24th of the last month after the most unpleasant journey that you can imagine, having been 40 hours between [the English port of] Dover and [the French port of] Calais with wind enough.[25]

This was a period when astronomy was flourishing throughout Europe, and Halley was eager to meet other members of the European astronomical community. In Paris, he met the celebrated Italian astronomer Giovanni Cassini (1625–1712), "my very particular good friend,"[26] who had moved there in 1669 to help set up the Paris Observatory, of which Cassini subsequently became the director. Among other achievements, Cassini was the person who discovered four moons of Saturn, discovered a gap in the rings of Saturn, measured the rotation rates of Jupiter and Mars, and noted that Jupiter was flattened at the poles.

It was while Halley was on his journey between Calais and Paris that he saw the return of the bright comet of 1680, "but the cloudy weather has permitted him to be but very seldom observed,"[27] and realized that comets were another subject for scientific research. (Halley's involvement with comets, in particular the one that was to be named after him, will be dealt with in chapter 3.)

Halley remained in Paris until mid-May. Then,

> practically always on the move, I pretty well made the circuit of France, and then passing into Italy, I penetrated as far as [Rome], tempted by the desire to see so great a city.[28]

His circuit of France included La Rochelle, Bordeaux, Toulouse, Narbonne, Montpelier, and Avignon.[29] He spent most of the rest of 1681 in Italy before returning to England in February 1682 via Genoa, Paris, and the Netherlands. His friend Robert Nelson fell in love with a young lady while in Rome, stayed there longer, and married her later that year.

MARRIAGE

In April 1682, at the age of 25, Halley married Mary Tooke,

> an agreeable young gentlewoman, and a person of real merit; she was his only wife, and with whom he lived very happily, and in great agreement, upwards of 55 years. He had by her, that lived to grow up, one son and two daughters; the son died before him, but the daughters are both living in great esteem, the one is single, and the other has been twice handsomely married.[30]

The wedding took place in St. James's Church on Duke's Place near Aldgate. We have no information on how they met or how long they had known each other before deciding to marry. One possible clue lies in the fact that St. James's was one where marriages could take place at very short notice with no need for the reading of banns beforehand. This could be taken as an indication of the desire for a hasty wedding. But, as with so much of Halley's personal life, we simply don't know.

We do know that the newly married couple, together with all Halley's astronomical equipment, moved to a house in the village of Islington, now a suburb of north London. Here he also began a series of observations of the Moon, whose position throughout time relative to the background of fixed stars did not line up well with predictions. Along with others, Halley realized that accurate knowledge of the Moon's position against this background would allow the determination of longitude at sea, and the aim of his observations was to help enable this. However, as the following chapters will show, his lifelong scientific explorations meant that he would spend relatively little time at home. His early French biographer, Jean-Jacques d'Ortous de Mairan, put it rather well:

> But neither domestic cares, not the tenderness of a happy marriage could diminish his ardor for the study of the heavens and the rest of nature, or confine him to his own country; we will still see him roaming the seas and bringing back new philosophical riches.[31]

2

PRINCIPIA

> *But for him [Halley], in all human probability, that work would not have been thought of, nor when thought of written, nor when written printed.*[1]

Isaac Newton's *Principia*, published in 1687, is universally accepted as one of the most important scientific texts ever written. It incorporated his theory of universal gravitation, together with his three laws of motion. Building on Johannes Kepler's three observationally determined laws of planetary motion, Galileo Galilei's terrestrial laws of motion, and the insights of Christiaan Huygens, René Descartes, and others, these simple mathematical laws enabled him to give an accurate description of how and why the planets move in their orbits, why an apple falling to the ground is obeying exactly the same laws as the Moon orbiting the Earth, why the oceans have tides, why comets behave as they do, why the equinoxes precess, why there are irregularities in the Moon's orbit, and much else besides. Their predictive power, the ultimate test for any scientific theory, was enormous. *Principia* brought about nothing less than a revolution in our view of the Universe.

1684: HALLEY, HOOKE, AND WREN

Yet it would probably never have been written were it not for a discussion that took place between three men one bitterly cold day in January 1684. The weather was so exceptionally cold that winter that the River Thames froze over for a couple of months and to such a depth that it was possible to safely hold frost fairs on the ice. Halley tells us that he journeyed from his home in Islington to central London one Wednesday,[2] probably for a meeting of the Royal Society.* Either before or after the meeting, for there is no record of their discussion in any of the Society minutes for January,[3] he met with two other Society members to confer on the seemingly obscure question of whether the consequence of an inverse square law of gravity could be that planets would move in ellipses.

(An inverse square law of gravity simply means that a planet twice as far away from the Sun as another would experience only one-quarter of the Sun's gravitational force, a planet three times as far away would experience only one-ninth of that force, and so on. An ellipse can be thought of as a squashed circle with well understood mathematical properties—this is explained in more detail below.)

The two other men were Sir Christopher Wren (mathematician, astronomer, a former president of the Royal Society, and the architect for the new St. Paul's Cathedral to replace the old cathedral that burned down in the Great Fire of London) and Robert Hooke (another prominent Society member and a brilliant experimenter whose considerable achievements would later be overshadowed by Newton's). Wren (1632–1723) was then in his early fifties; Hooke (1635–1703) was in his late forties. Halley, very much the junior at only 27 and having only joined

* There were three meetings of the Royal Society that January on Wednesdays—January 9, 16, and 23 (old-style Julian calendar)—and Halley was present at all three. No meeting was held on January 30 due to the anniversary of the death of King Charles I. So it is possible but unlikely that Halley could have been referring to January 2 or 30.

the Society a little over five years earlier, seems to have been recognized by his elders as a valued participant in their discussion.

All three accepted Johannes Kepler's observational conclusion of more than seventy years earlier that the planets moved around the Sun in ellipses, not circles. They also accepted Kepler's idea that this came about in some way because there was a force emanating from the Sun that weakened with distance. On the details of how this second point might work, Kepler's own speculations had been hopelessly wrong. Later ideas by others crystallized around the thought that the Sun exerted a gravitational force that reduced with the square of the distance—the inverse square law.

The problem was how to prove mathematically that this was the case. How do you demonstrate that Kepler's elliptical motion is one consequence of an inverse square law of gravitation? The demonstration itself is well beyond the scope of this book, but readers may find the brief background in the following paragraph useful.

CONIC SECTIONS

First of all, we need an understanding of all possible planetary orbits (and Newton would later demonstrate what these could be), including the ellipse. These go under the name of "conic sections." They were first described at length by the ancient Greek mathematician Apollonius of Perga, who had absolutely no idea that, nearly 2,000 years later, his discoveries would become so relevant to planetary motion. If you imagine slicing through a three-dimensional cone at various angles, you get in turn a circle, an ellipse, a parabola, and a hyperbola, as shown in figures 2.2 and 2.3. The first two are closed figures and so are plausible candidates for a constantly repeating planetary orbit; the second two are open and so are not. It is very easy to demonstrate mathematically that circular motion

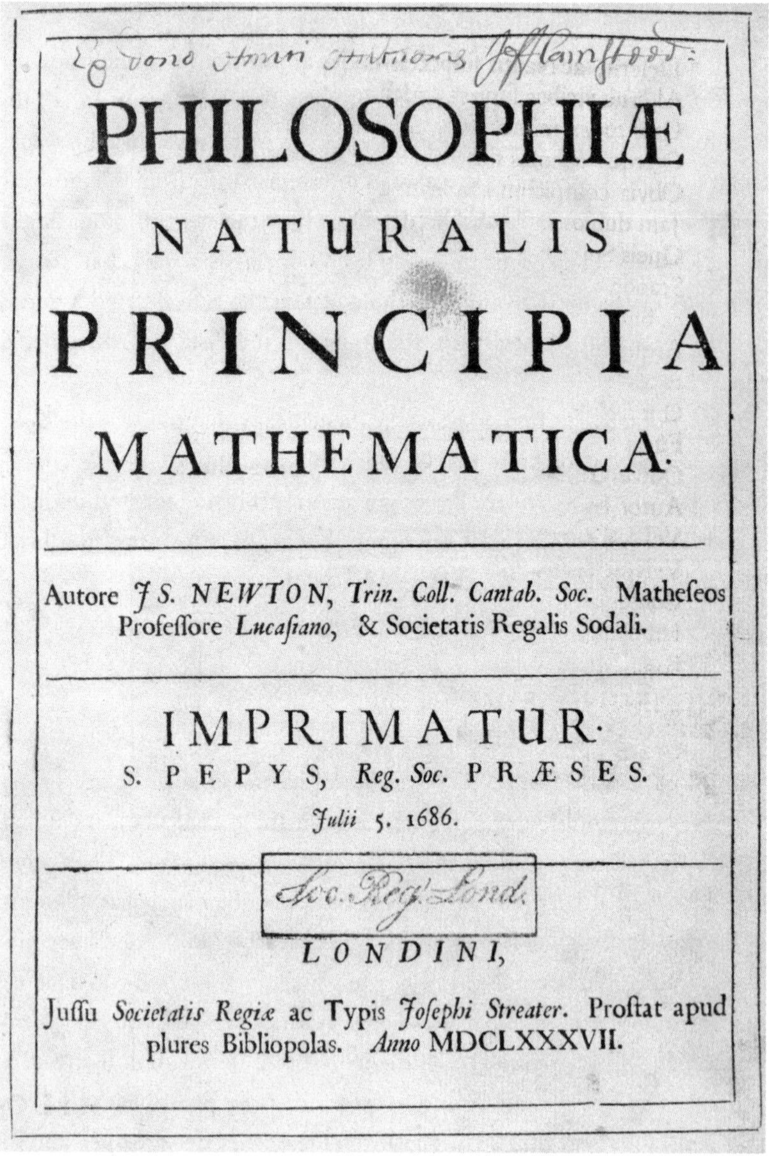
Ex dono Amici Authoris J Flamsteed:
PHILOSOPHIÆ
NATURALIS
PRINCIPIA
MATHEMATICA.
Autore J S. NEWTON, Trin. Coll. Cantab. Soc. Matheseos Professore Lucasiano, & Societatis Regalis Sodali.
IMPRIMATUR.
S. PEPYS, Reg. Soc. PRÆSES.
Julii 5. 1686.
Soc. Reg. Lond.
LONDINI,
Jussu Societatis Regiæ ac Typis Josephi Streater. Prostat apud plures Bibliopolas. Anno MDCLXXXVII.

Figure 2.1. The front cover of *Principia*. © *The Royal Society*

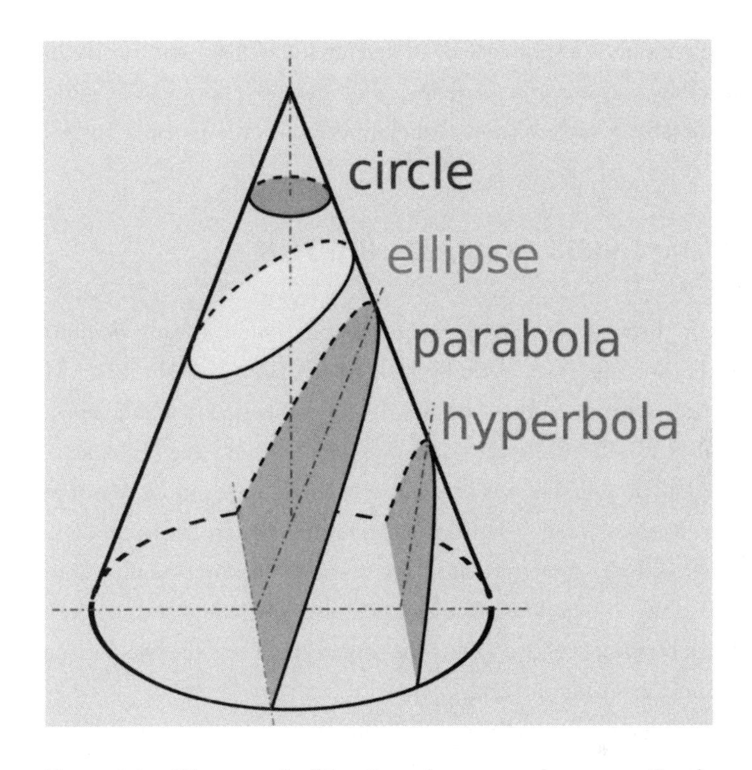

Figure 2.2. Diagram of slices through a cone, known as "conic sections." *Wikimedia Commons*

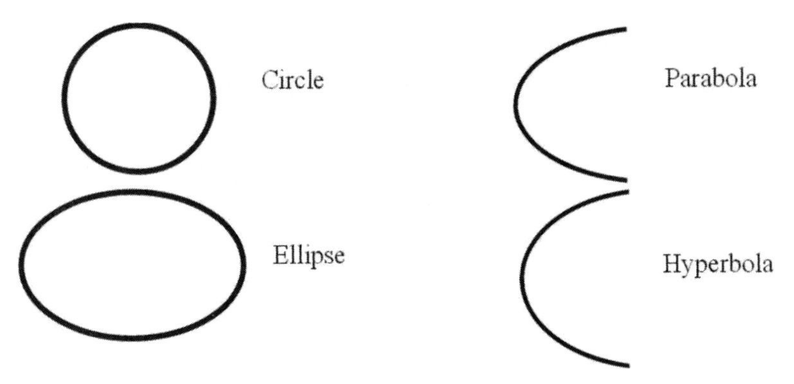

Figure 2.3. The four conic sections. *Created by author*

is one possible consequence of an inverse square law—for those who like math, this is set out in appendix E—but it was to take a mathematician of considerable genius to show that elliptical motion was also a possibility.

THE 1684 VISIT TO ISAAC NEWTON

In their discussion, Hooke claimed that he could already demonstrate that this was the case. Wren was skeptical and asked Hooke or Halley to supply a proof within two months in exchange for which Wren was prepared to offer the prize of a book worth 40 shillings. Hooke didn't manage to do this. He was certainly a talented man, but math just wasn't his strong point. Halley had already tried and failed to find a proof, but he was left with the thought that he knew someone who probably could—a still relatively unknown Cambridge professor of mathematics by the name of Isaac Newton (1642–1727). One source states that the two had already met once, probably in 1682.[4]

It is no exaggeration to say that, when his 1684 meeting with Newton finally took place, it would change the course of scientific history.

Halley would probably have visited Newton in late March or April of that year—after Wren's two-month deadline had expired and it had become clear that Hooke had failed to deliver the goods—were it not for a major catastrophe in his family. One morning in early March, his father left his house on Winchester Street in central London and was never seen alive again by his family. A reward of £100 was offered to anyone who should discover him, whether alive or dead. His body was finally discovered some six weeks later in April and some considerable distance away in Stroud near Rochester.

A boy, John Byers, who was walking along a riverbank in the area, found the body either in the river or on its banks. It had been stripped of all its clothes apart from shoes and stockings. He reported the discovery to others, one of whom—a Mr. Adams—had seen the advertisement offering the £100 reward. Adams traveled to London to inform Mrs. Halley. The body was subsequently identified by the dead man's nephew in spite of the fact that the face had been disfigured and thanks largely to the fact that the shoes and stockings were those of Halley's father. It was quickly recognized that the body had not been in the river for most or all of the time since he went missing, or it "would have been more corrupted." The conclusion of the court case was that Halley's father had been murdered, but there was also a suspicion that this was actually a suicide. Adams later tried to obtain the £100 reward for himself, and the case went to court. It was heard by the notorious Judge Jeffreys, who ruled that Adams should receive only £20 and that the balance of £80 should be paid into a trust for John Byers.[5]

Halley's father died intestate, and this led to a dispute between Halley and his stepmother Joane, who remarried relatively quickly,[6] as to how his father's assets should be divided. In June, a court arranged for a trust to be set up to divide those assets. Being involved in this kept Halley very occupied. He missed all the Royal Society meetings between March and October. So it is not at all surprising that his trip to Cambridge was delayed.

However, when Halley was finally able to visit Newton in August 1684, he hit the jackpot. As recounted by Newton's (and Halley's) friend, the French mathematician Abraham de Moivre (1667–1754),

> The Dr [i.e., Halley] asked him what he thought the Curve would be that would be described by the Planets supposing the force of attraction towards the Sun to be reciprocal to the square of their distance from it. Sir Isaac replied immediately that it would be an [ellipse], the Doctor struck with joy and amazement asked him how he knew it, why saith he I have calculated it.[7]

Halley immediately asked Newton to dig out the calculations. Newton looked through his papers and claimed that he couldn't find them but promised to locate them and send them on to Halley. Whether Newton did in fact have a valid proof at that stage or whether he was perhaps only playing for time is an interesting question. But it is a mark of his genius that he was able to produce one very shortly after this. Equally important, Halley had been able to persuade him to do so.

Now, however, came the tricky bit. Halley would shortly receive the proof and would immediately see the enormous significance of what Newton had done. He was keen to have it made public as soon as possible via the good offices of the Royal Society. But Newton was a notoriously prickly, sensitive, and withdrawn character who had already had two bruising encounters with Robert Hooke and was about to have another.

THE DISPUTES OF 1672 AND 1679

To understand the significance of these 1684 encounters with Hooke, we have to go back 12 years to 1672. Newton had just been made a member of the Royal Society on the strength of the reflecting telescope, which he had designed and built himself, the first time such a telescope had been constructed, and which had enormously impressed Society members. So impressed were they (and in order to try to secure priority rights) that they had sent a full description of the new invention to Christiaan Huygens, a Dutch citizen, widely viewed as the foremost European astronomer, physicist, and mathematician of his time.[8]

Newton had been so buoyed by the reception to his telescope that he had quickly followed it with a paper on optics (specifically, a paper on the nature of light and of color), for the first time publicly setting out his discoveries of the previous few years. This paper was to be published in its final form in 1704 under the title *Opticks* and is now regarded as one of Newton's most important works. Its initial reception, in spite of the new ideas that the paper contained, had been overwhelmingly favorable—but not quite totally so. Hooke, who not only was a prominent member of the Royal Society but also considered himself to be an expert in optics, had written a highly critical commentary on Newton's work.[9] Newton never forgave him and started to withdraw back into the shell from which he had only recently emerged.

A particularly significant exchange of letters between the two had occurred a few years later in 1679. Hooke, in his capacity as the recently appointed secretary of the Royal Society, had written to Newton to ask if he had any views on whether a planetary orbit could be mathematically constructed on the basis of an inverse square law of attraction between the planet and the Sun. A planet left to its own devices and not under any gravitational influence from the Sun or elsewhere would either be stationary or move in a straight line at constant speed. Could such motion in a straight line be mathematically combined with an inverse square law to produce a planetary orbit? Newton responded rather dismissively to this and didn't reply to Hooke's further letters.[10] However, it is arguable (and some people have argued[11]) that this was the first time that Newton had appreciated this way of looking at the problem and therefore that Hooke deserved some credit for bringing it to his attention.

DE MOTU: THE KERNEL OF *PRINCIPIA*

Following Halley's 1684 visit to Newton, things seemed to go smoothly, initially at least. To his great delight, in November 1684, Halley received from Newton (via Edward Paget) a nine-page paper titled "De motu corporum in gyrum" (On the motion of bodies in an orbit). This provided a mathematical proof that an inverse square law could result in an elliptical orbit (Kepler's first law) or could result in a parabolic or hyperbolic path for objects traveling over a certain speed (such as, perhaps, some comets, although neither Newton nor Halley appreciated that at this stage). It also derived mathematically Kepler's second law (that planets sweep out equal areas in equal times) and his third law (that for any planet, the cube of its distance from the Sun divided by the square of the time taken for a single orbit was a constant). It was the kernel of what would, over the next 18 months, become *Principia*.

Halley recognized that *De Motu* represented a huge step forward in celestial mechanics. Newton had shown that the three laws of planetary motion that Kepler had discovered and had thought of as separate and unrelated laws were in fact merely different manifestations of a single very simple law: the inverse square law of gravitational attraction. He soon made another trip to Cambridge to discuss this further with Newton. By December, he was able to report to the Royal Society as follows:

> Mr. Halley gave an account, that he had lately seen Mr. Newton at Cambridge, who had shewed him a curious treatise, De Motu; which, upon Mr. Halley's desire, was, he said, promised to be sent to the Society to be entered upon their register.[12]

There was now no stopping Newton. He seems to have realized that *De Motu* could be enormously expanded to include a whole range of phenomena. He spent the next 18 months, until the spring of 1686 (and beyond) working exclusively on this and abandoning all other activities, including, for example, his fascination with alchemy. Even the potential interruption of the death of Charles II in February 1685 and his replacement on the English throne by his (Catholic) brother James II did little to disturb his efforts.

INTERLUDE: HALLEY AS ROYAL SOCIETY CLERK

Meanwhile, Halley had been elected as clerk to the Royal Society at its January 27, 1686 meeting by 23 votes to a total of 15 votes for two other candidates for a salary of £50 per annum.[13] This new post had become necessary because the work of the Society had increased to such an extent that the honorary secretaries could no longer handle it on their own. His election was in spite of the fact that one requirement for the position had been that he should be single and have no children.[14] In fact, he had been married for nearly four years, and his first child, Margaret, seems to have arrived in 1685. One perhaps slightly odd requirement of the post was that Halley had to resign from his position as a Fellow of the Society (although he was to be reelected as a Fellow a few years later).

The post involved taking minutes of meetings, drawing up letters for one of the secretaries to sign, indexing Royal Society books, keeping a catalog of gifts, and dealing with correspondence with scientists both at home and abroad. Halley was also responsible for editing the Society's *Philosophical Transactions*, in which many of his own papers would be published. For at least some of these tasks, he was, of course, required to be fluent in English, French, and Latin. For taking minutes, he had to sit "at the lower end of the table; but to withdraw when particularly ordered."[15] Somebody clearly wanted to ensure that he did not get ideas above his station. He seems to have continued in this post until early 1698, presumably appointing a deputy during his absences.

It used to be suggested that, following his father's death, Halley was no longer wealthy and needed the income. However, extensive research by the late Sir Alan Cook has shown that Halley had sufficient investment income to remain comfortably off and didn't need this salary.[16] So the likely explanation for Halley's application for the post is simply that he wanted to involve himself more fully in the Royal Society.

It was just as well that he didn't need the money. The Royal Society had very little spare cash, having unwisely invested in a poorly selling book by the name of *A History of Fishes*, of which they now had a huge number of unwanted copies. So, rather late in the day at a meeting in July 1687, they decided that Halley should be paid his much-delayed first year's salary with 50 copies of *A History of Fishes*, with a gratuity of a further 20 copies, instead of £50. It is not recorded what Halley thought of this decision, but one can imagine.[17]

There was a further oddity concerning Halley's continuing position as clerk. At their June 1686 meeting, the Royal Society's Council minuted their satisfaction with their choice of Halley (in spite of the fact that, as they fully accepted, he was not able to satisfy their celibacy qualification[18]). But then in November, for reasons unknown and apparently in contradiction of their earlier resolution, it was decided that a new election should be held for the position of clerk "in the place of Mr HALLEY; and that it be put to the ballot, whether he be continued or not."[19] Then in January 1687, it was decided that "a committee [should] inspect the books of the Society, to see if Mr. HALLEY had performed his duty in relation to the entries to be made by him."[20]

Clearly somebody (or perhaps more than just one person) was trying to stir up trouble for Halley by trying to persuade Council members that he wasn't doing a good job. Perhaps it was Hooke, who may have felt that Halley was too readily taking Newton's side. Perhaps Royal Society members thought that Halley was spending too much time on *Principia*, rather than on Society matters. Most likely, it was John Flamsteed, the Astronomer Royal, who was trying to cause trouble for Halley. (The two had started off as good friends, but their completely different personalities had caused them to drift apart, and Flamsteed ended up greatly disliking Halley.) We don't know the reason, but the storm blew over in February, when it was reported that the books and papers of the Society were "in a very good condition," and Halley remained in post. It is a good thing that he did because otherwise *Principia* might never have seen the light of day.

PRINCIPIA EMERGES

By the spring of 1686, Books I and II of *Principia* were largely complete. At a Royal Society meeting on April 28,

> Dr Vincent presented to the Society a manuscript treatise intitled Philosophia Naturalis Principia Mathematica, and dedicated to the Society by Mr. ISAAC NEWTON, wherein he gives a mathematical demonstration of the Copernican hypothesis as proposed by KEPLER, and makes out all the phænomena of the celestial motions by the only supposition of a gravitation towards the center of the Sun decreasing as the squares of the distances therefrom reciprocally.
>
> It was ordered that a letter of thanks be written to Mr. NEWTON; and that the printing of his book be referred to the consideration of the council; and that in the meantime the book be put into the hands of Mr. HALLEY, to make a report thereof to the council.[21]

As a result, Halley effectively became proofreader for the book. And things continued to go smoothly for a while. At its May 19 meeting, the Society resolved that

> Mr. NEWTON's Philosophiae Naturalis Principia Mathematica be printed forthwith in quarto in a fair letter; and that a letter be written to him to signify the Society's resolution, and to desire his opinion as to the print, volume, cuts, etc.[22]

Then came the problem.

Hooke had seen the manuscript and was absolutely livid. In 1679, Hooke had made the suggestion to Newton that planetary ellipses could come about as a result of the planet's own motion combined with a force from the Sun that reduced with the square of the distance following an inverse square law. Yet Newton had given him no credit whatsoever for this. Hooke had often felt that people did not recognize the importance of his work, and this was the most egregious example to date. He made his feelings clear and was to remain bitter about this for the rest of his life.

Poor Halley now had to exert his diplomatic skills to the utmost. He wrote to Newton in May 1686 before Newton had the chance to hear an even less favorable account from anyone else. He started with flattery and praise:

> Your incomparable treatise, intitled Philosophiae Naturalis Principia Mathematica was by Dr. VINCENT presented to the Royal Society on the 28th past; and they were so very sensible of the great honour you have done them by your dedication, that they immediately ordered you their most hearty thanks.[23]

Then he broke the bad news:

> There is one thing more that I ought to inform you of, viz, that Mr Hook has some pretensions upon the invention of ye rule of the decrease of Gravity, being reciprocally as the squares of the distances from the Center. He [said] you had the notion from him, though he owns the Demonstration of the Curves generated thereby to be wholly your own; how much of this is so, you know best, as likewise what you have to do in this matter, only Mr Hook seems to expect you should make some mention of him, in the preface, which, it is possible, you may see reason to praefix.

Newton's anger took a while to build up, but after a few weeks, he too exploded in wrath. He informed Halley that he had already been fully aware of the likelihood of an inverse square law well before hearing about it from Hooke. And—as he pointed out—Hooke had merely made the suggestion, whereas he, Newton, had done all the hard work. Others (he meant Hooke) seemed incapable of doing the mathematical calculations.[24] He told Halley that he should check with Sir Christopher Wren, who would surely remember a conversation dating back to 1677 in which they had discussed the likelihood of an inverse square law. Wren would therefore be able to confirm that Newton already knew this. He was so annoyed at

Hooke's attitude that, to spite the world, he had decided to suppress the vitally important Book III of *Principia*, which he had so far not sent.[25]

So it was to take all Halley's considerable powers of persuasion, including establishing that Wren did remember the 1677 conversation, to finally calm Newton down.[26] Eventually, Book III duly appeared.

Sir Christopher Wren was an extremely capable mathematician and astronomer. Once Halley had explained the significance of *Principia* to him, he would undoubtedly have understood its importance and have been as eager as Halley to see it published. So it is at least possible that he said that he "remembered" the 1677 conversation purely to help Halley pacify Newton.

This was far from the end of Halley's problems. Unfortunately, and notwithstanding the resolution that was passed in May, the Royal Society was broke and could not afford the cost of getting *Principia* printed. So in June, its Council

> ordered that Mr. NEWTON's book be printed, and that Mr. HALLEY undertake the business of looking after it, and printing it as his own charge; which he engaged to do.[27]

In effect, Halley was now also becoming the publisher for the book as well as having to pay for it out of his own pocket. Even in April 1687, he was complaining that "the correction of the press costs me a great deal of time and pains."[28] He must have breathed a huge sigh of relief when, in July 1687 and entirely thanks to his efforts, he was able to write to Newton with the news that *Principia* had finally appeared in print.[29] He must also have been pleased that he, rather than the Royal Society president, was given the honor of presenting a copy of *Principia* to King James II.[30]

Halley's contribution to *Principia* was well summarized by Augustus De Morgan (1806–1871), when he said that

> this miracle of energy (for Halley is nothing less) occupied himself with the question of gravitation, sought for information from Hooke, Wren, and Newton, found out what the latter had done, induced him to begin the *Principia*, interested the Royal Society in its continuance, kept Newton up to his engagement, prevented him from mutilating it in disgust, undertook to see the work through the press, paid the expense of printing, and made himself thoroughly master of its contents, the most difficult task of all.
>
> But for him [Halley], in all human probability, that work would not have been thought of, nor when thought of written, nor when written printed.[31]

Halley was one of the relatively few people that Newton seems to have genuinely liked and respected, as can be seen by Newton's words in his preface to *Principia*:

> In the publication of this work the most acute and universally learned Mr. Edmond Halley not only assisted me with his pains in correcting the press and taking care of the schemes, but it was to his solicitations that its becoming public is owing; for when he had obtained of me my demonstrations of the figure of the celestial orbits, he continually pressed me to communicate the same to the Royal Society, who afterwards, by their kind encouragement and entreaties, engaged me to think of publishing them.[32]

This friendship was very much reciprocated by Halley. As late as 1725, for example, he wrote to Newton, calling him "the person in the Universe I most esteem."[33]

Principia is decidedly not a book for the fainthearted. It is deeply mathematical, mainly out of necessity but partly to annoy Hooke. It is also written in what a modern reader inevitably finds to be very archaic language. The first translation of the book from its original Latin into English in 1729 was by Andrew Motte. It ran to more than 400 pages of dense argument (in the modern edition published by Prometheus). It—and the concepts it embodied—dominated the physical sciences for the next 200 years, and Newton became a figure of legend. His and Halley's contemporary, the English poet, satirist, and supreme master of rhyming couplets Alexander Pope (1688–1744), wrote an appropriate verse:

> Nature and nature's laws lay hid in night:
> God said, Let Newton be! and all was light.

But nothing lasts forever. After some 200 years, small cracks began to appear in the Newtonian edifice. His formulae continue to be excellent approximations in most situations. However, we now know that the significantly more complex theories of special and general relativity formulated by the German physicist Albert Einstein (1879–1955) are a more accurate representation of reality. The counter-couplet from J. C. Squire provides an appropriate letdown:

> It did not last: the Devil howling "Ho!
> Let Einstein be!" restored the status quo.

3

THE COMET

Comets had, for hundreds and probably thousands of years, been thought of as harbingers of doom.[1] They were considered to be a warning of some terrible disaster that would occur in the near future and were certainly not objects that could be subjected to rational analysis. The famous lines of Calpurnia to her husband Julius Caesar in the eponymous play by William Shakespeare (1564–1616) summarized the standard view:

> When beggars die there are no comets seen;
> The heavens themselves blaze forth the death of princes.

This assessment was still widely held in Halley's time.

Aristotle had thought comets were entirely atmospheric phenomena.[2] The Danish observational astronomer Tycho Brahe (1546–1601) and others, such as the German astronomer Michael Maestlin (1550–1631), disproved this when they observed the bright comet of 1577. They noted that the comet displayed no daily parallax—in other words, for an observer on the rotating Earth, as the Earth's rotation carried the observer to the east, the position of the comet did not appear to move slightly to the west relative to the starry background. If the comet had been close by, it would have shifted its relative position. Therefore, the comet must have been traveling somewhere beyond the orbit of the Moon. This discovery does not seem to have influenced the general view that comets were a sign of troubles ahead.

Nor was this an unreasonable point of view for the times. The comet of 1066 had no doubt given warning of the Battle of Hastings, when William the Conqueror defeated King Harold in battle, ending the Anglo-Saxon rule of England. The bright comets of 1664 and 1665 had surely been warnings of the Great Plague of 1665 and the Great Fire of London of 1666.

Royal Society member and diarist John Evelyn (1620–1706), who we will meet again in a few chapters, tried to have it both ways when he said that comets "appear from natural causes." "Yet," he said, "they may be warnings from God, as they commonly are forerunners of his animadversions."[3]

Figure 3.1. Halley's Comet on March 8, 1986. ***NASA's NSSDC Photo Gallery***

His fellow Royal Society member Ralph Thoresby (1658–1725) had much the same view:

> Lord, fit us for whatever changes it [a comet] may portend; for, though I am not ignorant that such meteors proceed from natural causes, yet are they frequently also the presages of imminent calamities.[4]

As a more scientific approach slowly began to take hold, three different but related questions gradually emerged:

1. Were comets such as the two that appeared consecutively in late 1680 in fact one and the same, first approaching the Sun and then—a few weeks later—moving away from it?
2. Did comets ever return regularly after a period of many years, decades, or even centuries, or were they once-off phenomena?
3. Did comets obey laws of motion such as those that Kepler had found for the planets?

Kepler had failed to realize that comets moved in ellipses (or other conic sections) or that they obeyed his three laws of planetary motion just like the planets themselves.[5] He had instead thought that comets traveled in straight lines through the Solar System; later astronomers had similar views. But this was about to change.

Halley observed the bright comet of 1680 in November in the predawn skies as it moved toward the Sun while he was in London. He saw it again in December but this time in the evening skies moving away from the Sun, when he was traveling between the French port of Calais and Paris. The comet is now known as Comet Kirch after the German astronomer who first discovered it. Most people, including Halley at that time, had still not appreciated that this was one and the same comet. The minutes of Royal Society meetings at the time raise both possibilities.[6] The Astronomer Royal John Flamsteed did think that the two apparitions represented the same comet (and told Halley so), although Isaac Newton, at this early stage, did not.

But the Italian astronomer Giovanni Cassini, whom Halley had met in Paris, was ahead of the game in a different respect. He was already suggesting that comets were not once-off phenomena but returned time and time again. The two had been able to observe the comet together from the Paris Observatory.[7] Cassini explained to Halley that he thought the 1680 comet was the same as the one that Tycho Brahe had observed in 1577 and could also be the same as the one observed in 1665. Halley wrote rather uncertainly about this to Robert Hooke:

> I know you will with difficulty Embrace this Notion of his, but at the same tyme tis very remarkable that 3 comets should soe exactly trace the same path in the Heavens and with the same degrees of velocity.*

Cassini was wrong about these particular comets being the same, but he was absolutely right that the vast majority of comets are in orbit around the Sun and do return again and again although often only after hundreds or thousands of years. This was the first time this idea had been put to Halley, and he was later to become its most famous advocate.

There was another bright comet in 1682. The newly married Halley observed it from his home in Islington, not realizing that this comet would later immortalize his name.

As a result of his work on *Principia*, by 1685, Newton had also arrived at the conclusion that most comets were in closed elliptical orbits, meaning that they eventually returned. Specifically, he had also come to realize that Flamsteed was right and that the two apparitions of 1680, one moving toward the Sun and one later seen moving away, arose from the same comet. He had finally realized that all bodies in the Solar System,

* Letter to Robert Hooke (MacPike 1932, 51). In fact, Hooke had already made this very suggestion at a meeting of the Royal Society in March 1665 (and it was commented on by Samuel Pepys), but Halley (who would have been only seven years old at the time) could not be expected to have been aware of this.

not just the planets, were governed by his law of gravitation. He realized too that most comets orbited the Sun in ellipses, but he concluded that those ellipses were greatly extended so that their orbital periods (their "years") were much longer than those of the planets. (The longest planetary orbital period then known was Saturn's, at 29 Earth years.) As this was the case, their orbits were very close to being parabolas when close to the Sun and could be treated as such as a good first approximation,[8] which made the calculations significantly easier. And in 1687, the publication of *Principia* (with its detailed Book III, which included a section on cometary orbits and had nearly not seen the light of day) provided the mathematical means to demonstrate that Cassini was correct. Book III of the third and final edition of *Principia*—to be published in 1726—was to contain a large input from Halley on comets.

Cassini's suggestion had clearly been nagging at Halley as well, and he started to try to see if he could find actual examples of returning comets. By 1695, he found that he needed more data on past comets, in particular for the 1682 comet, to validate his growing suspicion that this comet had visited the Solar System on at least two previous occasions and so must be in an elliptical orbit around the Sun. However, his relationship with Flamsteed had now broken down to such an extent that he had to rely on the good offices of Isaac Newton to extract the vital data that would enable him to compute the orbit and compare it with the orbits of these earlier comets:

> I must entreat you to procure for me of Mr Flamsteed what he has observed of the comet of 1682 particularly in the month of September, for I am more and more confirmed that we have seen that comet now three times, since ye year 1531, he will not deny it you, though I know he will me.[9]

Newton was successful in getting the required information from Flamsteed, for which Halley was duly grateful:

> I give you many thanks for your communication of the observations of the comet of 1682.[10]

This enabled him to present to the Royal Society, in 1696, his initial conclusion that the comets of 1607 and 1682 were in fact one and the same. His further studies on comets were interrupted by other activities, including becoming deputy comptroller of the mint at Chester in the north of England and his voyages as captain of the *Paramore*, which will be tackled in chapter 6. However, he finally published a detailed paper on comets in 1705. The paper was written in Latin to ensure the widest possible circulation across Europe, but an English translation quickly became available. In the paper "A Synopsis of the Astronomy of Comets," he gave the orbital elements for 24 comets (dating back as far as 1337) for which he had examined the historical records. Three of these, the comets of 1531 (observed by the German astronomer Peter Apian), 1607 (observed by Kepler), and 1682 (observed by Halley from his Islington home), were separated in time by roughly 76 years.

Moreover, the detailed observations from those years enabled him to perform the huge number of extremely difficult calculations that showed that their orbits had very similar parameters. For example, all three orbited in the same plane, and the closest distance from the Sun was about the same for all three. He concluded that these were one and the same comet. The slight differences in the times to reach their closest distance from the Sun, he correctly believed, were due to the gravitational influence of the giant planets Jupiter and Saturn,[11] which could act to either speed up the comet or slow it down if it came too close to either of them. Although detailed observations were not available for the comet of 1456, the nearly 76-year gap convinced him that this was also the same comet. His final (and critical) conclusion was that

> I dare venture to foretell that it will return again in the year 1758.[12]

Halley died in 1742, but his prediction was spectacularly confirmed when the comet returned more or less on time. It was first observed on Christmas Day 1758 by an obscure German amateur astronomer and farmer named George Palitzsch.[13]

Halley had provided a perfect example of the increasing power of science to understand and predict the behavior of nature. After this, it became difficult to regard comets as anything other than completely natural phenomena that only sometimes (and entirely coincidentally) appeared before some great catastrophe. There was no longer any need to invoke the wrath of a divine being to account for the appearance of a comet.

More specifically, Halley had demonstrated the enormous predictive power of the science set out in Newton's *Principia*. After the reappearance of his comet, any remaining doubts about Newton's achievement evaporated.

Shortly after its reappearance, the French astronomer Nicolas-Louis de Lacaille, in 1759, was the first to call it "Halley's Comet," and this well-deserved name has stuck. We now know that Halley's Comet has been observed several times in the past. For example, the comet of 1066, mentioned earlier, which makes an appearance in the Bayeux Tapestry, was also Halley's Comet. It also seems certain that the appearance of Halley's Comet in 1301 was the inspiration for the comet that is shown as the Star of Bethlehem in *The Adoration of the Magi*, painted by Giotto (ca. 1267–1337) in the early 1300s.

However, these are by no means the earliest sightings of Halley's Comet. The earliest confirmed records of sightings are from China, referring to its appearance in 240 BCE, although a comet observed in both China and Greece in 467 BCE may also have been Halley's. In 1986, Halley's Comet returned to the inner Solar System for the first time since the start of the space age. It was met by a small flotilla of craft from the United States, the Soviet Union, Europe, and Japan. It was entirely appropriate that the spacecraft launched by the European Space Agency

Figure 3.2. Halley's Comet can be seen in the Bayeux Tapestry. *Wikimedia Commons*

was named *Giotto* after the painter who had observed the comet more than 600 years earlier. *Giotto* approached the comet to within 600 kilometers, closer than any of the rest of the little flotilla. It discovered that the comet's nucleus was dark and irregular, shaped rather like a peanut (or, if you prefer, a potato), and about 15 kilometers long and 10 kilometers wide. It consisted of solid silicates and carbon, with numerous openings that allowed volatile gases (water vapor, carbon monoxide, and carbon dioxide) to escape from underneath.[14] The original description by the astronomer Fred Whipple (1906–2004) of a comet as a "dirty snowball," a mixture of ice and dust, is not far off the mark. (It has been suggested by some that "icy dirtball" might be a closer description in the light of the findings of the *Giotto* mission.[15])

We now know far more about comets in general than was known in Halley's time. They are deep-frozen time capsules that preserve a record of conditions that existed during the formation of the Solar System more than 4.5 *billion* years ago. They are believed to originate from one of two large reservoirs of material in the outer Solar System. Shorter-period comets usually come from the Kuiper Belt, named after the Dutch astronomer Gerard Kuiper (1905–1973), and consist of debris in orbit beyond the orbit of Neptune. Longer-period comets are believed to originate in the Oort Cloud, named after another Dutch astronomer, Jan Oort (1900–1992), which is a vast, very distant spherical cloud of comets surrounding the Solar System, perhaps extending halfway to the nearest star. From time to time, gravitational perturbations cause some of these to be dislodged from their otherwise roughly circular orbits and fall in toward the Sun. Halley's Comet, although it has a relatively short period of only 76 years, is nevertheless believed to originate from the Oort Cloud[16] because it orbits in the opposite direction from all the planets and most other short-period comets.

The nucleus of a comet is a mixture of volatile ices, such as those of frozen water, carbon monoxide, and carbon dioxide, together with more solid material. As a comet gets closer to the Sun, it warms up, and these volatiles can escape as gases, forming its dust tail. (The word "comet" is derived from the Greek word *kometes*, which means "long haired," an appropriate description for a comet.) In fact, comets have two tails: a dust tail and an electrically charged ("ion") tail, which is less visible. The ion tail appears first. The volatiles outgassing from the comet's nucleus are stripped of electrons ("ionized") by the solar wind and are then pushed away from the nucleus by that wind as it streams from the Sun. This tail always points directly away from the Sun and exhibits a blue color.

The dust tail is formed as some of the solid material of the comet's nucleus breaks away from the warming nucleus. The heavier dust particles are not so affected by the solar wind as the ionized gases and gradually drift into a curved orbit. Eventually, the dust particles that have been shed by some comets form the regular meteor showers seen on the Earth as we pass through the comet's orbit. Halley's Comet itself has given rise to two such meteor showers—the Orionids in April and the Eta Aquarids in October—caused by the Earth's passage through the dust orbit twice a year. So even though the comet itself has disappeared to the outer reaches of the Solar System, the debris remains strung out along the comet's orbit, and the debris that hits the Earth's atmosphere is seen as a meteor shower.

Dust tails shine white in reflected sunlight and can be extremely long. Examples are known that would stretch from the Earth all the way to the Sun. More than 4,000 comets are known, but this is believed to be a small fraction of the total number of comets in existence. All comets gradually lose mass each time they get close to the Sun. So they do gradually get smaller and smaller and eventually disappear, although Halley's Comet will still be around for people to watch for some time to come.

At the time of publication of this book, Halley's Comet is getting very close to its aphelion (farthest distance from the Sun), some 35 times the Earth-Sun distance, and somewhere between the orbits of Neptune and Pluto. As these words roll off the printing press, it is about to begin its long fall back toward the Sun.

Its next perihelion (closest approach to the Sun) will be on July 28, 2061, just over 300 years since the return predicted by Halley and almost exactly 100 years after the first human spaceflight—that of Yuri Gagarin. It may not be too fanciful to imagine that on this return, Halley's Comet could have human visitors.

4

THE ROYAL SOCIETY

Many readers will already have been familiar with the fact that Halley was the indispensable driving force behind the creation and publication of *Principia* and that he was the first person ever to predict the return of a comet, one that was subsequently named after him. However, one of the key themes of this book is that he achieved so much more than just these two things; these later accomplishments will be described in this and subsequent chapters. His major contribution to our understanding of geomagnetism deserves a chapter of its own and will be dealt with in chapter 6. His equally critical role in celestial distance measurement will be tackled in chapters 7 and 9.

In addition, Halley contributed numerous papers to the Royal Society during the late 1680s and 1690s (once his work on the publication of *Principia* was largely complete) on an astonishingly wide variety of other topics that went well beyond his interest in astronomy. These were published in the Society's *Philosophical Transactions*.

This chapter looks at just seven of the more noteworthy of these in that period out of a total of more than 50. They show that Halley was far more than merely the man who first discovered a periodic comet or merely the man who was the midwife to *Principia*.

Even though he is still thought of primarily as an astronomer, of the seven areas discussed below, five have no connection at all with astronomy, and the connections of the remaining two are relatively tenuous. They demonstrate above all that he possessed in abundant quantities that essential characteristic of a good scientist: a profound sense of curiosity.

THE TRADE WINDS (1686)[1]

Halley wrote a lengthy paper for the Royal Society in 1686 in which he explored the causes of the trade winds. These are the winds that generally blow throughout the year from northeast to southwest in the region between about 30° north and the equator, with mirror-image winds blow-

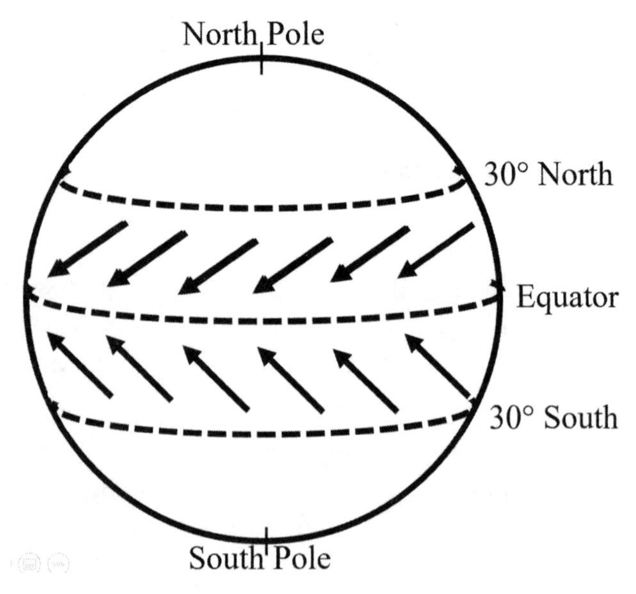

Figure 4.1. The trade winds. ***Created by author***

ing from southeast to northwest in the region between about 30° south and the equator. These latter winds—as his paper pointed out—he had experienced at firsthand and, to his great chagrin, during his stay on St. Helena. Navigation was becoming ever more important with the growth of shipping, so an understanding of atmospheric circulation had come to be seen as increasingly desirable.

Halley's map of the trade winds (figure 4.2) was the first of its kind and an important contribution to the subject.

Halley correctly recognized that, because the Sun's heat is greatest in the equatorial regions, this will cause the air there to expand and rise. Air from adjacent regions will therefore inevitably flow in from north and south of the equator, creating the trade winds. He also correctly deduced that the trade winds involved circulation of the air so that, for example, high above the winds that are traveling from northeast to southwest, there must be equal and roughly opposite winds traveling from south to north.

However, Halley was incorrect in his explanation for the east-west component of the trade winds. He argued that this was because the Sun (which from the Earth's point of view travels westward in the sky every day) causes the hottest point on the equator to always move westward. So in his view, the east-west component is indirectly caused by the Earth's rotation, but the direct cause is heating by the Sun.

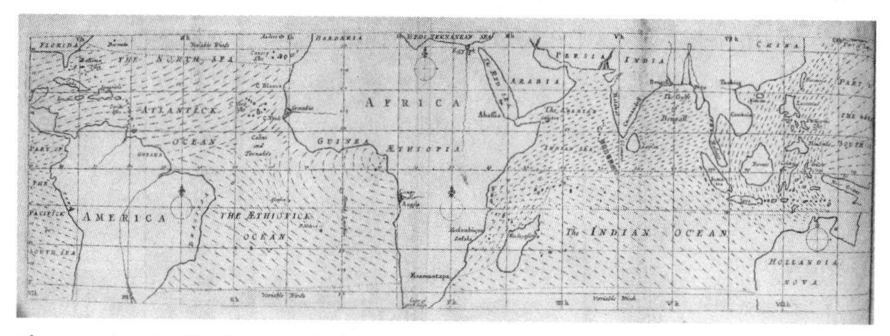

Figure 4.2. Halley's map of the trade winds. It is often considered to be the first meteorological chart. © *The Royal Society*

The correct explanation[2] for this east-west component is now known as the Coriolis effect. It was first established by George Hadley, nearly 50 years later. In 1735, Hadley tied this component directly both to the Earth's rotation and to the distance that the air has to move away from the Earth's axis, although it took time for his explanation to be accepted. This was in part because some people muddled "Hadley" with "Halley." A full discussion of the Coriolis effect is outside the scope of this book, but an explanation of how it causes the trade winds is given in appendix G.

EVAPORATION AND WATER CIRCULATION ON THE EARTH (1687)[3]

The water on planet Earth is constantly being recycled. Evaporation from the oceans is offset by rain, falling either directly back into the oceans or onto land—and hence via rivers back to the sea. In 1687, Halley performed an experiment to determine the rate of evaporation of water. He applied this rate of loss to the Mediterranean and compared it with his estimates of water arriving from the largest rivers that flowed into the Mediterranean to check whether the two balanced out.

To calculate the evaporation rate of water, Halley took a pan of water about 10 centimeters deep and 20 centimeters in diameter and heated the water to the temperature of a hot summer day. The pan was attached to some scales so that the weight of evaporated water could be measured. Knowing the density of water, he could then calculate the volume of water that evaporated in a given time. Using a series of very reasonable assumptions, he arrived at a figure for the daily evaporation for the whole of the Mediterranean.

To determine the rate of arrival of water in the Mediterranean, Halley first carried out measurements of the rate of flow of water along the River

Thames at Kingston Bridge in London. He assumed that the nine largest rivers* flowing into the Mediterranean each carried about 10 times as much water as the Thames. This was a deliberate overestimate but was his way of allowing for the numerous very small rivers that also flowed into the Mediterranean. This enabled him to calculate a figure for the volume of water flowing into the Mediterranean.

Halley's calculations showed that only just over one-third of the water evaporated from the Mediterranean was returned via the rivers flowing into it. However, given both the large uncertainties and the significant assumptions in his calculations, he did well to find that the figures were so close. His paper concluded by noting that water flowed into the Mediterranean from the Atlantic, with the implication that this could account for at least some of the difference but that further research was needed.

1688: THE YEAR OF THE GLORIOUS REVOLUTION

As Halley interacted with royalty throughout his life, it should be mentioned that 1688 (the year after Halley's paper on evaporation) was the year of the Glorious Revolution, when (the Catholic) King James II was deposed and went into exile and was replaced on the throne by his (Protestant) elder daughter, Mary, and her husband, William III of Orange, from the Netherlands. Halley nearly got himself into trouble following the changeover, as he continued to express his admiration for both Charles and James because of the various favors he had received from them. When William heard of this, he was "a little alarmed" but was persuaded that Halley's admiration was purely a result of gratitude and did not mean that Halley agreed with their religious views, in particular those of James. Halley will again be coming into contact with William in chapter 6.

* The nine rivers were the Rhone, the Tiber, the Po, the Danube, the Ebro, the Dniester, the Borysthenes (Dnieper), the Tanais (Don), and the Nile.

The year 1688 is also when the second of Halley's two surviving daughters, Catherine, was born. Given the high level of infant mortality in those times, it is possible that other children were born to the Halleys either before or after Catherine but did not survive—we simply don't know.

DIVING BELL (1691)

In a paper to the Royal Society in 1689, Halley explained his plans for building a diving bell. In 1691, he built one and was able to send people—including himself—below the water level in it, initially in the River Thames and then at Pagham on the south coast of England, where an English frigate, the *Guynie*, had recently sunk. As can be seen from figure 4.3, the diving bell was in the shape of a truncated cone attached at the bottom to three heavy weights to keep it in an upright position. At the top, Halley had made a window to provide the occupants with some light. There was a device at the top to remove stale air, and new air was supplied in barrels sent down separately. There was even a handy shelf that the occupants could sit on. On August 26, he

> read an account of his late experiment of the diving bell, wherein he had maintained 3 men 1¾ hours [underwater] in ten fathoms deep [about 18 meters, or 60 feet] and showed by what method he got the air down into the bell. Which he proposed as practicable for any depth, and number of men required.[4]

One month later, he further reported to the Royal Society that

> he had contrived a way to go out of the diving bell, and stay in the water as long as he pleased, and be at liberty to do what he pleased there, by a vessel a man may carry on his head, like a cap, and by forcing to him with bellows, or such like means, the air contained in the diving bell.[5]

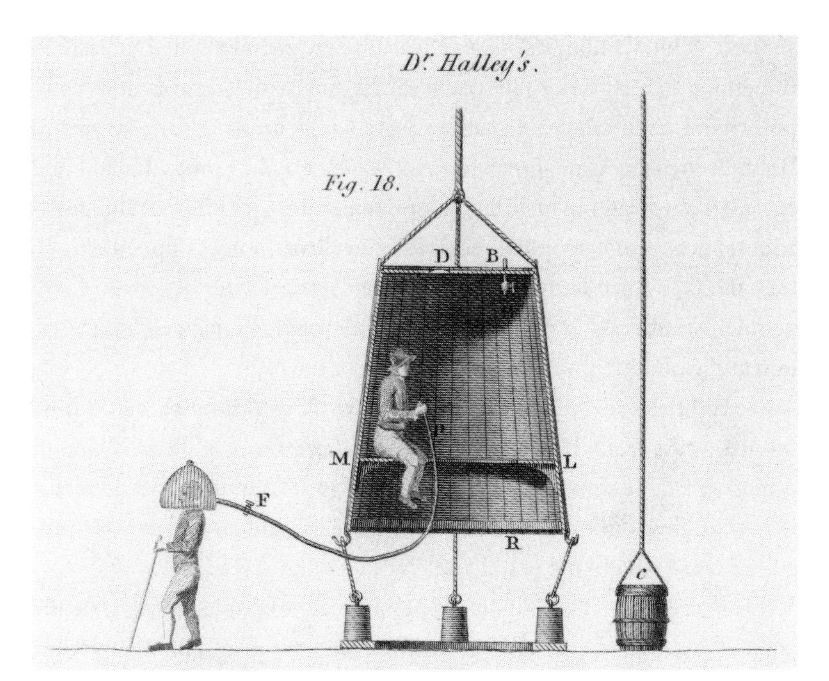

Figure 4.3. Halley's diving bell. ***Science Museum/Science and Society Picture Library***

Sure enough, the diagram also shows (bottom left) Halley's diving suit, complete with a flexible pipe back to the diving bell so that the diver could be supplied with fresh air. Halley was not the first person to construct and use a diving bell, but his construction seems to have been an improvement over previous versions.[6]

LIFE EXPECTANCY TABLES (1692–1693)[7]

Halley can claim credit for being one of the first people to construct life expectancy tables based on actual population data; in effect, he was one of the founders of actuarial science. He was not the first to look at this subject; John Graunt, another Royal Society member, had attempted something similar some 30 years earlier but with arguably less comprehensive data.[8] Halley's analyses were based on data from the city of Breslau, then the capital of Silesia in Bohemia (and now in Poland and renamed Wroclaw), which had kept detailed records of both the births and the deaths of its population and—critically—the ages of people when they died. It was a fairly isolated and stable population of some 34,000 people. Numbers of births and deaths were roughly equal, and immigration and emigration were negligible.

In 1692, Halley acquired detailed records of deaths by age during the five years from 1687 to 1691. From these records, he was able to construct a table showing how many people out of an initial 1,238 births would be alive one year, two years, three years, and so on later. He presented these data to the Royal Society.*

From the table, it was therefore possible, for example, to calculate the probability, at any age, of death within the next year, although Halley put this only second on his list of uses. He saw other possibilities as well. For example, he said that one use of the table would be to calculate the proportion of the population available for military service, defined as the number of males between the ages of 18 and 56, which turned out to be just over one-quarter of the total population.

Halley also saw the table as a means of calculating life expectancy at any age. For example, there were 531 people in his list aged 30. There would come a time in the future when half of these (currently) 30-year-olds would still be alive but the other half would have died from a variety of causes. Half of 531 (after rounding down) is 265. By consulting his table, he could see that 265 people would still be alive at a point in time

* Halley's table, shown on page 59, is slightly confusing because the "Age 1" figure of 1,000 is what he decided was the average number of people in their first year of life, that is, people who had not yet reached the age of one. He also seems to have rounded some of his numbers. The figure of 1,000 seems to have been derived from the average of the number of live births of 1,238 and the number still alive at the age of one year of 890, which is 1,064 and which he has rounded to 1,000. So, strictly, "Age 1" is actually "Age ½."

Halley's table of Breslau people living by age:

Age	Number	Age	Number	Age	Number	Age	Number	Age	Number	Age	Number
1	1,000	8	680	15	628	22	585	29	539	36	481
2	855	9	670	16	622	23	579	30	531	37	472
3	798	10	661	17	616	24	573	31	523	38	463
4	760	11	653	18	610	25	567	32	515	39	454
5	732	12	646	19	604	26	560	33	507	40	445
6	710	13	640	20	598	27	553	34	499	41	436
7	692	14	634	21	592	28	546	35	490	42	427
Age	Number	Age	Number	Age	Number	Age	Number	Age	Number	Age	Number
43	417	50	346	57	272	64	202	71	131	78	58
44	407	51	335	58	262	65	192	72	120	79	49
45	397	52	324	59	252	66	182	73	109	80	41
46	387	53	313	60	242	67	172	74	98	81	34
47	377	54	302	61	232	68	162	75	88	82	28
48	367	55	292	62	222	69	152	76	78	83	23
49	357	56	282	63	212	70	142	77	68	84	20

between ages 57 and 58, so this was the average life expectancy of all 30-year-olds in Breslau. (The modern definition is slightly different and somewhat more complicated.) This in turn provided a means of calculating the price of insurance policies and annuity payments.[9]

THE AREA OF ENGLAND AND WALES (1693)

An example of Halley's sheer ingenuity was in his method, published in early 1693, for establishing an accurate figure for the total surface area of England and Wales at the request of one John Houghton, another Royal Society member. (Scotland was at that time still a separate country.) He took the best map of England and Wales then available and carefully cut off the sea around the country's irregular coastline. He then cut out the largest circle possible from the map. Knowing the scale of the map, he could easily calculate the actual area represented by the circle. He then simply weighed both the cutout circle and the map of England and Wales as a whole. Then, by the law of proportions, the weight of the circle to the weight of the whole map will be the same as the area represented by the circle to the area represented by the map. Problem solved. His answer, at 38.66 million acres, was not very different from the modern figure of 37.3 million acres.[10]

THE FLOOD (1694)[11]

The book of Genesis in the Old Testament tells the story of how God had come to regret having created humanity and had decided to wipe out almost the whole human race as well as virtually all the animals by drowning them in a flood that covered the entire Earth. The only people he allowed to be saved were Noah and his family. Under God's instructions, Noah built a boat that was also sufficiently large to take two of every kind of animal so that the surviving humans and animals could start life again once the flood had subsided.

Since the flood is stated to have covered the whole Earth, one has to assume that animals from the other side of the planet, such as koalas and kangaroos, were also rescued, although no details are given in Genesis as to how they were collected or how they were returned to their normal habitats when the flood subsided.

This story was still universally accepted as true in Halley's time. The only issue at stake was whether the flood was a miraculous event or whether it had occurred naturally. Halley's view, expressed in a paper to the Royal Society in 1694, was that the flood could be explained in entirely natural terms. His calculations showed that the 40 days and nights of rain that Genesis said had occurred would be insufficient to cover more than low-lying land. So he speculated that the flood had come about because of the gravitational effect of a comet that had approached very close to the Earth.

We now know that the gravitational effect of a comet on the Earth, no matter how closely it approached, would be negligible, so Halley's explanation would not work. Nevertheless, it provides another example of his inventive mind.

Interestingly, although the paper was presented in 1694, it was not formally published in the Royal Society's *Philosophical Transactions* until 30 years later, in 1724, when Halley's position in scientific society had become unassailable. A note at the end of the publication indicates that the delay was because the author feared that he might "incur the censure of the Sacred Order." In other words, a purely natural explanation for the flood might have offended the Church at the time.

SNOWDON (1697)[12]

The brilliant Italian scientist Evangelista Torricelli (1608–1647), a student of Galileo, is most famous for his invention of the barometer. The idea behind the barometer is quite simple. A horizontal glass tube, sealed at one end, is completely filled with the liquid metal mercury and then moved into a vertical position with the unsealed end in a pool of mercury. The height of the mercury then gives a direct measure of atmospheric pressure. (The space created in the tube above the mercury is a vacuum.) If you then take the barometer up a mountain, the height of mercury will drop because the atmosphere becomes thinner and exerts less pressure.

In a paper to the Royal Society as far back as 1686, Halley had established that an average figure for the height of mercury at sea level was about 76 centimeters (30 inches). He had also noted that the fall in the height of mercury with increasing altitude could be used as a means of measuring altitude above sea level and that the height of the mercury also fell as rain approached and rose as better weather came.[13] (That pressure drops with altitude is the operating principle of aircraft altimeters to this day.) His friend John Caswell had already used a barometer to establish that the height of Snowdon, the highest mountain in Wales, was about 1,134 meters (tolerably close to the modern figure of 1,085 meters).[14]

In May 1697, Halley had an opportunity to experiment further in this area. In March 1696, at the behest of Isaac Newton, who had just moved to the Royal Mint in London to take control of a currency at that time in crisis, Halley moved to the town of Chester in the northwest of England. The problem that needed tackling was the debasement of the currency that had resulted from the widespread practice of clipping a small amount from coins and then using the clippings to create new coins. The chosen way around the problem was to issue only coins with milled edges so that any clipping could be immediately detected. But this meant the withdrawal and replacement of all coins without milled edges. This was a mammoth and tedious task, but there was no realistic alternative.

Halley was in Chester for about two years as the deputy comptroller of the branch of the Mint in that city at a salary of £90 per annum. Chester was one of five temporary regional centers set up outside London to help deal with the recoinage. Newton clearly wanted somebody he could trust in the job, although Halley didn't enjoy the experience, which he said was "at best . . . but drudgery, but as we are in perpetual feuds is intolerable."[15] He was at least able to take advantage of an eclipse of the Moon to enable a calculation of Chester's longitude to be made.

The opportunity for further experimentation with a barometer came because Snowdon is not unreasonably distant from Chester. So Halley decided to repeat the experiment performed by John Caswell. He and a group of people he had assembled[16] climbed Snowdon with his barometer, where he found that the height of the mercury was only 66.3 centimeters compared with his sea-level measurement the following day of 75.9 centimeters. This gave a figure for the height of Snowdon that was much the same as Caswell's.

After sending a paper on this (as well as other papers from Chester) to the Royal Society and following his return to London, Halley's prodigious output fell off for a while as he set off on his voyages around the Atlantic Ocean, to be described in chapter 6.

5

HALLEY'S CHARACTER AND RELIGIOUS SKEPTICISM

What was Edmond Halley like as a person? As well as having obvious intelligence, he was someone who was apparently full of boundless self-confidence and energy. If he had not had these qualities, he could never have produced such a huge volume of original work. He also seems to have been liked by most people (with one very notable exception). It is arguable that our sources are a little sycophantic in their praise, but it is worth quoting a few of them. *Biographia Britannica* says about him that

> he was of a middle stature, inclining to tallness, of a thin habit of body, and a fair complexion, and always spoke as well as acted with an uncommon degree of sprightliness and vivacity. . . . He . . . possessed all the qualifications necessary to please princes who were desirous of instruction, great extent of knowledge, and a constant presence of mind; his answers were ready, and at the same time pertinent, judicious, polite and sincere.[1]

In his Éloge, the French biographer Jean-Jacques d'Ortous de Mairan says that

> the reputation of others didn't worry him, . . . jealousy had no place in his heart. He also had no time for these outrageous prejudices in favor of just one nation that are so injurious to the rest of the human race. Friend, compatriot and disciple of Newton, he nevertheless spoke of Descartes with respect. . . . To conclude, these qualities that are both rare and worthy of respect were enhanced by a profound joyfulness that neither old age nor the paralysis that he suffered a few years before his death was ever able to change.[2]

A later source says that he

> was open-hearted, generous and quick to lend a hand. He was . . . much more enthusiastic about the success of Newton's *Principia* than regarding anything that he himself accomplished, yet his own many achievements made him the "second most illustrious of [English] philosophers."[3]

Annoyingly, we have virtually no information on Halley's personal and family life, both before and after his marriage to Mary Tooke, beyond the basic family tree set out in appendix A, and even this has gaps. It is frustrating that there is a lack of personal (as opposed to scientific) correspondence between Halley and others. There are also no letters between him and his wife and no indication at all as to whether she traveled with him—to Chester, perhaps, or to Oxford. There is a statement in the biography believed to be by Martin Folkes that

> she [Mary] was his only wife, and with whom he lived very happily and in great agreement, upwards of 55 years.[4]

But a cynic might argue that Halley's absences from home were so frequent and so lengthy that there was little time for the two of them to quarrel. Even when at home, he must have spent huge amounts of time writing up his next scientific paper. He was throughout his life the ultimate workaholic.

There is also an inscription on Halley's tombstone, now on display at the Royal Greenwich Observatory, written by his two daughters:

> This stone was consecrated to excellent parents by two devoted daughters in the year 1742

His one real enemy was Flamsteed. The initial relationship was warm and friendly. At the time Flamsteed was just setting out as the first Astronomer Royal, he had welcomed the help of a bright and enthusiastic (and unpaid) assistant. While he was still a student at Oxford University, Halley had helped Flamsteed on a number of occasions. In September 1675, the two of them observed an eclipse of the Moon together, and in June 1676, they jointly observed an eclipse of the Sun.[5]

But the friendly relations between the two degenerated irrevocably, probably in the mid-1680s. Nobody knows exactly why. It must have been in part because of their very different personalities. Flamsteed was a narrowly religious man. He was also an obsessive with not the least trace of a sense of humor. He could not have been more unlike the broad-minded, gregarious, and lighthearted Halley. It can't have helped that Flamsteed had health problems (about which he frequently complained) throughout his life, whereas Halley was somebody fortunate enough to have had abundant health and energy.

Flamsteed's letters from the mid-1690s on are overflowing with vitriolic complaints about Halley. The dislike seems to have been one-sided. On one occasion, in 1703, Halley asked Flamsteed what he could do to repair the broken relationship between them. Flamsteed replied that "he must become a just, serious and virtuous man."[6] And in a letter written by Halley to Flamsteed in 1711, Halley asked Flamsteed to

> govern your passion, and when you have seen and considered what I have done for you, you may perhaps think I deserve at your hands a much better treatment than you for a long time have been pleased to bestow on [me].[7]

It is certainly the case that Flamsteed suspected Halley of plagiarism: "his art of filching from other people, and making their works his own."[8]

Specifically, he thought that Halley had stolen ideas on the magnetism of the Earth from a math teacher, Peter Perkins,[9] but there doesn't seem to be any evidence that this was the case. (For Halley's contribution to our understanding of geomagnetism, see chapter 6.) There is a real possibility that Halley borrowed without acknowledgment James Gregory's idea of using a transit of Venus to determine the distance to the Sun, although Flamsteed doesn't seem to have picked this up. (Halley's pivotal role in this distance determination is outlined in chapters 7 and 9.) The fact that Flamsteed didn't do so tends to argue that both he and Halley were unaware of Gregory's idea. Flamsteed also seems to have doubted whether Halley's religious beliefs were sufficiently orthodox: "I pray God give him grace to see his follies and repent,"[10] and "I pray God convert him."[11] He also commented on "him [Halley] and his infidel companions."[12]

On the question of religious orthodoxy, he may well have been correct. Halley certainly acquired a reputation for broad views on questions related to religion. Exactly how broad his views were has been a matter of considerable debate and something of a minefield that one enters at one's peril. Lengthy and learned academic papers have been written for the Royal Society on the subject by Simon Schaffer and Dmitri Levitin.[13]

Eugene F. MacPike, who nearly a century ago made a huge study of Halley's life and work, concluded that "there is no doubt that Halley did hold liberal views on religion and was very outspoken concerning the subject."[14] For example, he quoted the story of a Scottish person who paid several visits to a coffee shop known to be frequented by Halley in order to "see the man that has less religion than Dr [David] Gregory."[15] MacPike adds that Halley was also described as "a *very* free thinker."[16]

David Gregory himself, the nephew of James, was hardly a model of propriety. While in his previous post at Edinburgh University, he was variously accused of several misdemeanors, such as not taking the Lord's Supper, attending church with his pupils only on the first Sunday of the session, swearing, drunkenness, pugnacity, fondness for women, superficial teaching, and taking too long a holiday at Christmas. So Halley must have been thought to be even worse than this.[17]

Further back in time, François Arago, who in 1855 wrote a brief biography of Halley, said that "Halley, comme l'assurent tous ses contemporains, portait le scepticisme jusqu'à ses dernières limites" (Halley, as all his contemporaries assure us, carried skepticism to its utmost limits).[18]

It has also been both widely believed[19] and strongly disputed that the target of Bishop George Berkeley's book *The Analyst: or A Discourse Addressed to an Infidel Mathematician* (published in 1734, when an elderly Halley was still alive) was in fact Halley. Freethinkers at the time had argued that there was a huge contrast between the clearheaded logic of mathematics and the muddled and imprecise thinking of theology. The book tried to demonstrate that mathematics also lacked logic in, for example, the calculus that Newton and Leibnitz had independently devised and therefore was just as bad as theology. However, if the book was not aimed at Halley, it is difficult to know who else it could have been directed at. Another proposed candidate was Newton, but this seems unlikely, as Newton was very devout—albeit unorthodox—in his religious views and was in any case dead by the time that *The Analyst* was published. In addition, the very fact that many people accepted that the book had (or could have had) Halley in its sights indicates that he was widely thought of as a dangerous freethinker.

1691

Matters came to a head in 1691, when the post of Savilian Professor of Astronomy[20] at Oxford University became vacant. The person leaving the job, Edward Bernard, was departing for a better-paid post as a cleric but was to die only five years later. Halley, along with two other candidates (David Gregory and John Caswell), put in applications. Halley had the backing of the Royal Society, but Gregory had Newton's backing. This does not indicate any hostility toward Halley on the part of Newton, merely that he thought Halley would be more useful elsewhere. It sounds strange to modern ears, but the decision on whom should be selected for this purely scientific post rested in part with the Archbishop of Canterbury, John Tillotson.[21] Later accounts plausibly state that Bishop Edward Stillingfleet (Bishop of Worcester) and his chaplain, Richard Bentley, were also involved.

According to one later account, part of the selection process involved an interview with Stillingfleet, who was clearly asking awkward theological questions that Halley did not want to be forced to answer. He therefore tried to change the subject by responding

> My Lord, that is not the business I came about. I declare myself a Christian and hope to be treated as such.[22]

Another report on Halley's suspected broad religious views came from William Whiston, who was a theologian and mathematician and a contemporary of Halley's. His report may not be reliable because of his own unorthodox religious beliefs. However, he gives an account of a meeting between Halley and Dr. Bentley in which Halley had been "so sincere in his infidelity that he would not so much as pretend to believe the Christian religion, though he was thereby likely to lose the professorship."[23]

Halley didn't get the job, even though there is a strong case for saying that he was the best candidate. It is therefore not unreasonable to suppose that his failure must have been because of his assumed lack of orthodoxy. Those who argue otherwise have failed to come up with a credible alternative reason. Instead, the job went to his friend David Gregory.

Some have questioned the reliability of some of these stories because they appeared only much later. In particular, in 1844, the Reverend S. J. Rigaud wrote a vigorous and indignant pamphlet titled *A Defence of Halley against the Charge of Religious Infidelity*. He pointed out that

Whiston, for example, was writing about 55 years after the event and in any case had his own axe to grind. Interestingly, Rigaud did accept that Halley was indeed the target of Berkeley's *The Analyst*, even though he strongly questioned the validity of Berkeley's assumed sources.

However, at least a couple of sources are contemporary. Nicolas Fatio de Duiller, in a comment (quoted in the 2013 Royal Society paper by Dmitri Levitin) in a letter to Christiaan Huygens only a few months after the events of 1691, said that Halley had lost "à cause des opinions qu'on lui a voulu imputer sur la Religion" (because of the opinions which people wished to attribute to him on religion). And before Halley had even been interviewed for the job, the outgoing Savilian Professor of Astronomy at Oxford, Edward Bernard, gave his view (again quoted by Levitin) that Halley would have difficulties in getting it because of the "complaint of several persons . . . touching ye irreligion & even blasphemous discourses of Mr Halley." These earlier examples make the later ones more credible.

Part of the problem is that Halley himself has left few clues as to his real feelings.* There is no doubt that in any conflict between a scientific finding and a religious belief based on the Bible, Halley would take the side of the former and simply "reinterpret" what the Bible said to make it compatible with the science. The clearest example of this was when his analysis of the saltiness of the oceans (to be discussed in chapter 7) caused him to realize that the age of the Earth was probably considerably greater than the biblically derived figure of about 6,000 years. To get around the problem, he simply redefined the biblical word "day" (which in Genesis clearly means a period of 24 hours) to mean instead a very long period of time. We now know that the Earth is more than 4.5

* It may be worth noting, however, that his *Commonplace Book* (consisting of jottings he made mostly while at St. Paul's or as a student at Oxford) shows that he had become aware of a wide range of religious beliefs that were both sincerely held and at odds with the beliefs of the Church of England. See the Royal Society paper "The Commonplace Book of Edmond Halley" by Edward H. Cohen and John S. Ross, published in 1985.

billion years old rather than the mere 6,000 or so years implied in the Bible and widely accepted in Halley's time. Halley was certainly heading in the right direction.

Rather than believe in miracles, Halley also sought natural explanations of events that the Bible claimed had happened. So, instead of assuming a miraculous source for all the extra water required for the biblical flood to cover the Earth to a considerable depth (as discussed in chapter 4), he concluded that the gravitational effect of the close approach of a comet had somehow caused the flood. This did not in itself necessarily qualify him for the title of "infidel"; many devout Christians of the time were attempting to make their beliefs seem more reasonable by ascribing natural causes to apparently miraculous events. Nevertheless, it is arguable that this reading of the Bible in a more academic and critical way (rather than in a merely devotional way) is one of the factors that has led to a slow decline in religious belief—and Halley was one of the people who initiated this approach. Whether he realized that this would be one result of this tactic is impossible to tell.

(Interestingly, although the jury is still out on exactly where all the water in the Earth's oceans originally came from, it is increasingly likely that this was at least in part the result of bombardment of the Earth by comets and—mainly—asteroids, billions of years ago.)

One area in which Halley was determined to show his orthodoxy was in what is arguably a relatively peripheral area known as the "eternity of the world." Orthodox Christian belief demanded that the Earth and, by extension, the Universe have both a beginning (as stated in Genesis) and an end (as required by some parts of the New Testament). Halley was anxious to show, both before and after his failure to get the post at Oxford, that he accepted both these propositions. However, he held these beliefs only on what he saw as scientific grounds. (For example, in a paper he presented to the Royal Society in 1692, he argued that the ether, which he believed was necessary to enable light to propagate through space, would also, over time, have a tiny effect in slowing down the Earth in its orbit. This would eventually and inevitably lead to the Earth slowly spiraling in toward the Sun. He therefore concluded that the Earth could not always have been in its current orbit and would not stay in its current orbit forever.)

It seems unlikely that all the stories told about Halley are completely without foundation. Such tales didn't make an appearance for many leading natural philosophers of the day (although there are a few examples) and one is left wondering why they should have done so for Halley unless there was some justification for them.

Equally, however, Halley was certainly not an atheist. Atheism at that time was difficult to sustain, and it is undoubtedly the case that Halley believed in the existence of a God. One of his presentations to the Royal Society, for example, referred to "the wisdom of the Creator,"[24] although such references by him are not very common. He may, perhaps, best be described as a variety of deist: someone who believed that a Creator God was necessary to bring the Universe and its laws into being but who did not intervene further after the initial act of creation.

But whatever Halley's precise views were, he was a man of his time. He lived in a period when some form of religious faith was still very much the default position and died more than a century before the publication of Charles Darwin's *Origin of Species* drove a coach and horses through the argument from design (for the existence of God). It was to be even longer before astronomers became fully aware of the utter insignificance of the Earth in a Universe that (if the theory of the multiverse is correct) looks more like a product of random forces than of design. Given the change over time in the balance of scientific evidence, a hypothetical modern-day Halley might well be an atheist,[25] but Halley himself was something other than this. What exactly his real views were we shall probably never know.

6

HALLEY, PETER THE GREAT, AND THE VOYAGES OF THE *PARAMORE*

We do not often think of him as a sailor; and yet, previous to [Captain James] Cook, Captain E Halley was our first scientific voyager.[1]

Peter the Great was a larger-than-life character. He was czar of Russia from 1682 at the somewhat early age of 10 until his death in 1725 at the age of 52. He was responsible for introducing sweeping reforms that aimed to westernize and modernize Russia. He spent more than a year on a diplomatic mission to western Europe, aimed in part at gaining knowledge of shipbuilding techniques in order to build up and improve his navy. His visit included just over three months in England during early 1698 at about the time that Halley was returning to London from his post at the Mint in Chester. Peter himself was endeavoring to travel incognito as part of a small group of Russians to avoid having to attend formal events. This attempt at anonymity fooled nobody, however, as his great height (slightly over two meters, or about six feet eight inches) made him easily recognizable.

Peter amply achieved his aim of learning about shipbuilding techniques from the English, who thereby inadvertently helped to launch

Russia on the long trajectory toward becoming a military superpower. He took back to Russia both knowledge and skilled English manpower to train his own countrymen. His own skills at steering ships did, however, leave something to be desired. On two separate occasions when Peter was commanding vessels sailing on the Thames, there were collisions that caused considerable damage.[2]

Notoriously, while in London, Peter acquired a mistress, Letitia Cross, who was a singer and actor. On his departure, Peter paid her 500 guineas for her services, but she didn't consider this enough.[3]

We know that Peter met Flamsteed at the Royal Greenwich Observatory on four occasions, sometimes while Flamsteed was carrying out observations, as this was recorded in passing in Flamsteed's notes.[4] The account in *Biographia Britannica* tells us that, also while in London, Peter met Halley:

> When Peter the Great, Emperor of Russia, came into England, he sent for Mr Halley, and found him equal to the great character he had heard of him. He asked him many questions concerning the fleet which he intended to build, the sciences and arts which he wished to introduce into his dominions, and a thousand other subjects which his unbounded curiosity suggested; he was so well satisfied with Mr Halley's answers, and so pleased with his conversation, that he admitted him familiarly to his table, and ranked him among the number of his friends.[5]

We cannot be certain of the source of this information, but it was apparently written within a few years of Halley's death, and there seems no reason to doubt its overall accuracy. There is a similar account in the earliest biography of Halley, written (probably) by Martin Folkes, even sooner after Halley's death[6] and another (again similar) account in the (admittedly partly derivative) eulogy[7] by Jean-Jacques d'Ortous de Mairan.* A

* It is unlikely but not impossible that all three of these sources about Peter the Great had a single common origin and that this origin is not correct. It is certainly the case that the biography by Mairan is partly but by no means entirely a derivative of the Folkes biography.

few historians have questioned whether the two did in fact meet because of a lack of immediately contemporary records. It would be surprising if they were right given Halley's increasing prominence in English society and the fact that he had the scientific and nautical knowledge that Peter needed. There are several days of Peter's visit when there is no record of what he did or whom he saw,[8] making one or more meetings perfectly credible.

Figure 6.1. The statue of Peter the Great in Deptford, overlooking the Thames. ***Photo courtesy of the author***

At the time, Peter and his entourage had rented out a house in Deptford (southeast London) that was owned by Sir John Evelyn and that was conveniently close both to the dockyards and to the Royal Greenwich Observatory. There is still a Czar Street in Deptford to commemorate his visit. Unfortunately, their appalling behavior caused massive damage to both the house and the garden. Evelyn was mortified by the damage, although he did receive compensation from the state of more than £350.

The destruction in the garden probably provided the origin of the story that—following a heavy drinking session one evening—Halley pushed Peter round the garden in a wheelbarrow. This included pushing him through John Evelyn's beloved yew hedge.[9] The story is unlikely to be true if only because its origins cannot be established, but it does encapsulate something of the characters of both men.

THE *PARAMORE*[*]

There is also one piece of evidence that indicates that Peter may have sailed in the *Paramore*, a ship shortly to be used by Halley. A letter from the English Admiralty dated March 16, 1698, authorizes the preparation of the *Paramore* for Peter's use: "The Czar of Muscovy having desired that his Majesty's Pink the Paramore at Deptford may be rigged and brought afloat, in order to make some experiment about her sailing."[10] This can also be seen as a further indication that Peter and Halley met; given that Halley was about to set sail in the *Paramore* himself, he would have been the ideal person to show off the ship to Peter.

Peter finally left England, to the huge relief of his hosts, on April 25, 1698. About six months later, the *Paramore* set sail from Deptford, with Halley as commander and a total complement of 20 men.[11] The voyage

* As was usual for spellings in those times, a number of different spellings were used for the *Paramore*. I have adopted the one used most by Halley.

had been planned for many years, but events had conspired to prevent sailing until now. It was certainly extremely unusual for Halley, a landsman with no naval career behind him, to be appointed to the post of captain, and it can be taken only as an indication of the high regard in which Halley was held by those who appointed him.

Halley's son (also Edmond), his third child, was probably born sometime during 1698, but it is an indication of our lack of personal information on Halley that we simply do not know whether his son was born before or after he had set sail on the *Paramore.*

Marine navigation was of increasing importance to England's burgeoning maritime interests in the seventeenth and eighteenth centuries. Safe, efficient, and timely voyaging could not be achieved without it. To navigate effectively, mariners needed to know both where they were and in which direction they were heading. Once out of sight of land, this became much more difficult. Two issues were particularly problematic—first, determining the ship's longitude and, second, determining its direction relative to true north, which in turn required knowledge of compass variation, explained below. It was the latter issue that Halley's voyage was to address, although he would also have some opportunity to determine the longitudes of some places on land that they visited. The purpose of the voyage, commissioned by King William III, was stated as being "to improve the knowledge of the Longitude and variations of the compass."[12]

COMPASS VARIATION

Navigational charts are based on true north, but the mariners of Halley's time often simply determined the direction of north using a magnetic compass. Unfortunately, there is a difference between true north and magnetic north, known as compass variation (sometimes called declination).

To navigate accurately, it is vital to be able to correct for this variation wherever you are.

True north—as shown on a map or a chart—is the direction to the North Pole, which is the point where the imaginary straight line around which the Earth rotates every 24 hours passes through the surface of the Earth. This is in the Arctic (and there is an equivalent South Pole in the Antarctic). For almost all practical purposes, the position of the North Pole on the Earth's surface is fixed. In fact, it does change very slowly but only by tiny amounts. It has moved only some 20 meters since 1900.

Magnetic north is the direction toward which a compass needle points. The needle of a perfect compass will align itself with the Earth's magnetic field wherever it is put. This magnetic field is created by electric currents in the Earth's molten iron-nickel core, some 2,900 kilometers below the surface. The shape of the field is complicated, but in general it looks like that of a gigantic (and entirely imaginary) bar magnet tilted relative to the Earth's axis of rotation. The magnetic North Pole is where the Earth's magnetic field is perpendicular to the Earth's surface.

To complicate matters further, the imaginary bar magnet moves by significant amounts over time. The magnetic North Pole is believed to have been somewhere in North America in Halley's day. It was first located in 1831 by James Clark Ross in the Canadian Arctic, but it has since passed close to the true North Pole, has moved beyond it, and is now heading toward Siberia at about 55 kilometers per year.

Halley's voyages were intended to systematically plot the size of the variation at a large number of points over the Atlantic Ocean. Knowledge of the degree of variation with position both enables more accurate navigation and gives clues as to the nature of the Earth's magnetic field.

LONGITUDE

The longitude problem, as it was known, was a separate issue. Position on the Earth's surface is defined by two coordinates: latitude and longitude. Latitude is the number of degrees north or south of the equator. Longitude is the number of degrees east or west of a north–south line passing through Greenwich, England.

As explained in chapter 1, it is relatively easy to determine latitude at sea by measuring the angle above the horizon of the Sun at noon or of the Pole Star at night. Determining longitude accurately, however, requires a knowledge of both the local time and the time at Greenwich. The difference in the two times effectively gives the longitude. A one-hour time difference equates to a 15° longitude difference, a two-hour time difference to a 30° longitude difference, and so on.

There were, in principle, only two ways of establishing this difference accurately. The first was for a ship to carry a clock giving the local time at Greenwich, but this was not then possible. The best clocks of the time were pendulum clocks, but these quickly went wrong when on board a moving ship. (The pendulums simply could not cope with a ship's motion.) Alternatively, it had to make use of a "clock in the sky" by, for example, noting the Moon's position relative to background stars. Knowing what that position would be at Greenwich would again enable longitude to be calculated. However, it was still not possible to predict the precise position of the Moon at any moment in the future, and the exact positions of large numbers of background stars were still being laboriously determined by Flamsteed.

In other words, neither method of determining longitude at sea was remotely adequate at that time. Instead, seamen often had to rely on the (unreliable) estimate of longitude by the dead-reckoning method. This involved working out a ship's new position by estimating its speed and direction of motion away from an initial known position.

In Halley's day, accurate longitudes had not necessarily been determined even for some fixed sites on land. One of his achievements on these voyages was to establish the longitudes of some of the places where the ships landed. Doing this was possible, for example, by waiting for a lunar eclipse and then measuring the exact time of the beginning (or end) of the eclipse. All observers, no matter where they are, see the eclipse start (and finish) at the same moment and also know the local time at which the event occurs. So a subsequent comparison of two local times, both at the place whose longitude is to be determined and at Greenwich, enables longitude to be determined. The same method is possible using eclipses of the moons of Jupiter as they pass behind the planet; these are also seen at the same moment by all observers no matter where they are. Halley was able to use both methods to determine the longitudes of various ports during his voyages.

THE VOYAGES: CAPTAIN HALLEY

Halley had been interested in the subject of the Earth's magnetism for a long time. Even in his schooldays, in 1672, he had carried out his first measurement of magnetic variation before leaving for Oxford.[13] He had found that, at that time in London, there was a difference of 2½° between magnetic and true north. The specific context of the voyages lay in a paper that he had presented to the Royal Society in 1683. In it, he had listed observations showing that there were large differences in the amount of variation by both time and place. In some places, the magnetic North Pole appeared to the west of the true North Pole and in other places to the east. Having to his own satisfaction demolished the theories of others, he put forward the novel idea that—to account for the differences in variation—there are four magnetic poles (each with different powers of attraction) inside the Earth: two near the true North Pole and two near the South Pole. This idea is yet another example of his inventive mind but is no longer accepted; two magnetic poles are now usually used to describe—if not to explain—the observations. Toward the end of his paper, there is the gentlest of hints that more observations of the variation are needed, thereby laying the groundwork for the voyages of the *Paramore*.

Halley's first voyage, beginning in October 1698, was almost entirely in the North Atlantic via Madeira, the Cape Verde Islands, Paraiba (Brazil), Barbados, and Antigua. The route of the voyage is shown in figure 6.2. Only a single voyage should have been needed, but this one was far from being a total success. On arriving at the Cape Verde Islands, Halley's ship was fired on by two English ships whose captains had initially taken him for a pirate. Fortunately, no damage was done.

The lack of wind meant that progress was slow for several weeks after that, and Halley was compelled to land in Paraiba to take on new supplies of fresh water. Here, an eclipse of the Moon enabled him to determine the longitude of that location. When he later arrived at Barbados, he was able to use observations of the times of the eclipses of Jupiter's satellites to calculate the longitude of the island.

During the voyage, there was also some resentment among certain crew members that Halley was in command of the vessel in spite of the fact that he had not previously been in the Royal Navy so had not exactly followed a conventional route to the post of commander. While the ship was sailing close to Brazil, his boatswain seems to have deliberately set an incorrect course in order to miss an island that Halley was keen to map. Halley discovered this in time to correct the course, but worse was to follow. Later in the voyage, his lieutenant, Edward Harrison, openly and in front of several other crew members questioned his ability to command the vessel. Underlying his hostility seems to have been resentment that,

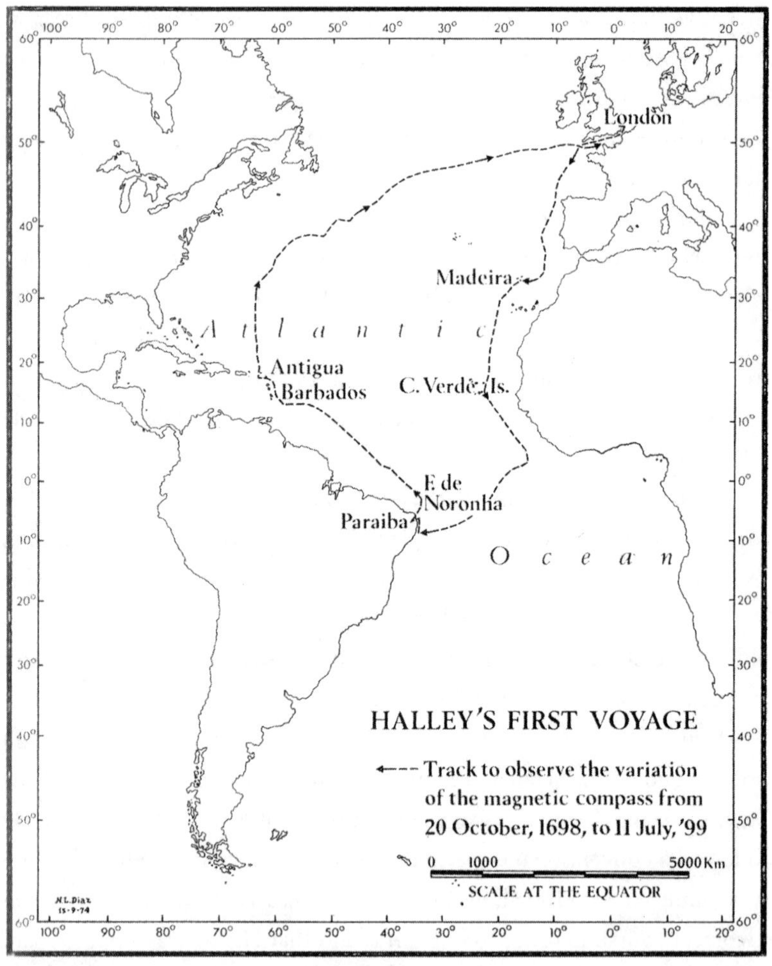

Figure 6.2. Halley's first voyage. "The Three Voyages of Edmond Halley in the Paramore." ***David Higham Associates***

some four years previously, Halley had been critical of a book written by him.[14] The book had proposed some impractical methods for determining longitude at sea. Harrison also disobeyed direct instructions from Halley, who concluded that he had no choice but to take charge of navigation himself and return to England to have Harrison and various other crew members replaced. His attempts to get the recalcitrant members of his crew court-martialed on return were a failure. Instead, to his great disappointment, the court simply reprimanded them, saying that there may have been some grumbling among them, as there generally was in small vessels under such circumstances.

As a result of these problems, the voyage lasted only about eight months and came to an end in July 1699. Nevertheless, Halley did manage to make about 50 measurements of magnetic variation before returning.

Halley's second—and more successful—voyage, with some new crew members (including a boatswain with only one arm), began in September 1699, lasted almost a year, and covered much more of the Atlantic (see figure 6.3).

After initially following much the same route as before, Halley docked at Rio de Janeiro, in part to get supplies of rum for his crew, then proceeded southward. While in the South Atlantic, he sailed far enough south to encounter icebergs. There was a perilous moment in the voyage, when "we were in imminent danger to [lose] our ship and lives, being [surrounded] with ice on all sides in a fog so thick that we could not see it till was ready to strike against it."[15] He made one of his relatively rare errors in later describing these to the Royal Society as fixed islands.[16]

Returning northward and away from the icebergs, the ship stopped at St. Helena, where the water they had hoped to renew their supplies with was "so thick with a brackish mud that when settled it was scarce fit to be drunk."[17] From here, via another island that did have good drinking water, Halley sailed to Pernambuco (Recife) in Brazil, where a self-important Englishman claiming to be the local consul, one Mr. Hardwyck,

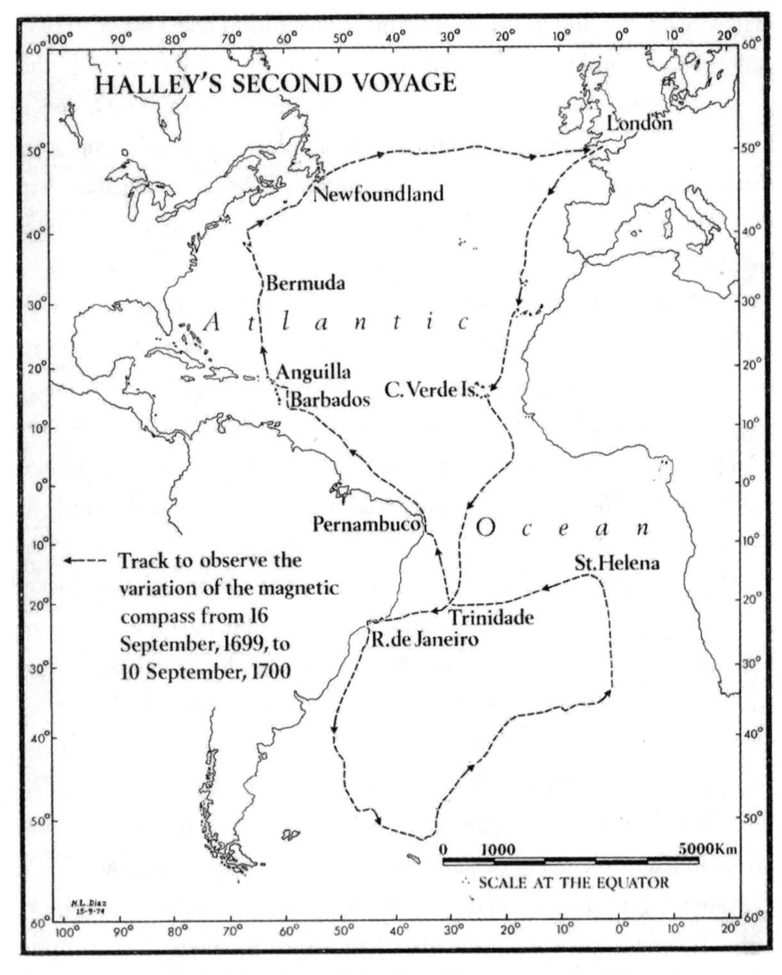

Figure 6.3. Halley's second voyage. "The Three Voyages of Edmond Halley in the Paramore." *David Higham Associates*

was another person who initially decided that Halley must be a pirate. Notwithstanding the fact that Halley was able to show him his commission papers, he kept Halley under guard in his house until he had carried out a personal inspection of the ship.

From here, Halley sailed roughly northwest and in 16 days arrived at Barbados. Here he "found the island afflicted with a severe pestilential disease." Even though he himself stayed on the island for only three days, he and many of his men were struck down with it. We don't know what the disease was, but yellow fever, typhoid, or some form of gastrointestinal problem brought to Barbados with the slaves have all been suggested. This was one of the relatively few occasions in his life when Halley became unwell. Fortunately, everyone recovered, thanks in part to "the extraordinary care of my doctor." From here, they sailed home via Antigua, Bermuda, and Newfoundland, arriving back in Deptford in September 1700.

On this second voyage, Halley was able to make roughly 100 measurements[18] of magnetic variation and to establish the longitude of a few places on land. Both voyages were notable by the fact that he lost only a single crew member, a rarity for such expeditions. It happened when a cabin boy, Manley White, fell overboard in stormy weather near Madeira on the second voyage and could not be rescued. This was something he always regretted and "never mentioned it without tears."[19]

Halley's observations now enabled him to publish early in 1701 the first map of magnetic variation over the whole of the Atlantic Ocean (see figure 6.4).* It was arguably also the first such map to show the data by drawing lines along which there were equal values of magnetic variation, now known as isolines or isogones (and sometimes Halleyan lines). This

* In principle (although, sadly, not in practice), it would have been possible to determine a ship's longitude using this map. A seaman simply had to determine both latitude and compass variation at his location, then consult Halley's map to find the longitude of the place that had the two values he had determined. In practice, the method would work only where the lines of constant magnetic variation ran roughly north–south.

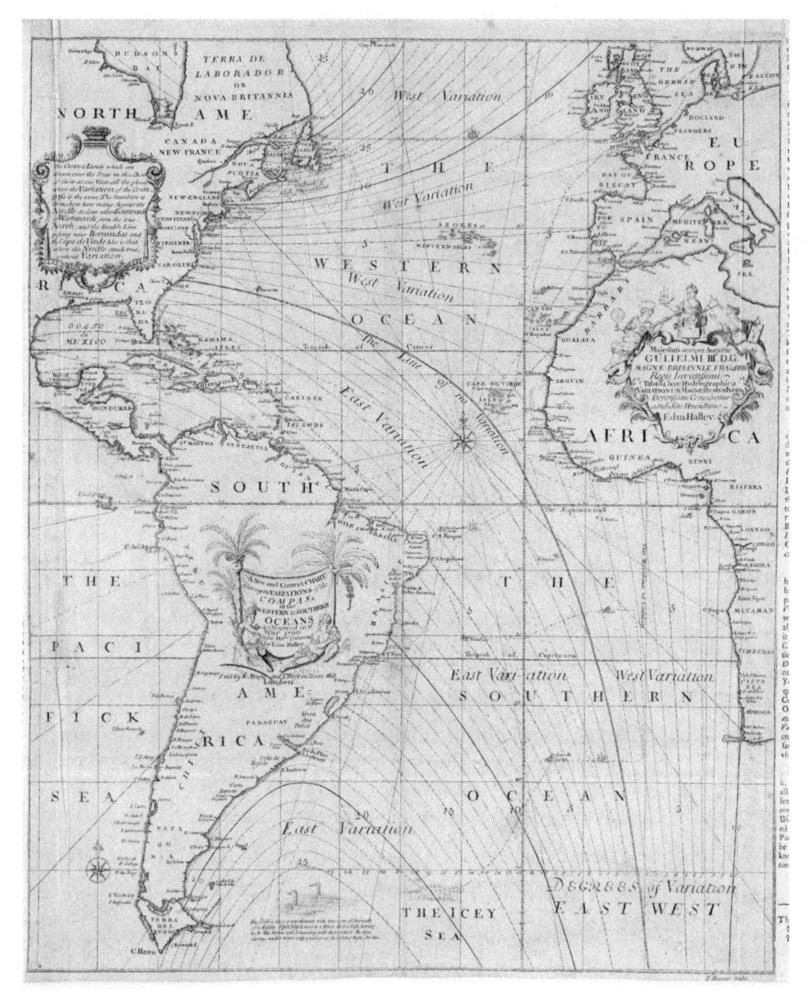

Figure 6.4. Halley's map of magnetic variation in the Atlantic. ***New York Public Library Digital Collections***

might now seem an obvious way to show the data, but it was probably original on Halley's part. His efforts did not go unrecognized or unrewarded. King William III ordered that he be paid the sum of £200 for his work in addition to his pay as a naval captain.[20]

Halley's scientific voyages were not quite yet at an end. He asked for and was promptly given permission to undertake a study of the tides in the English Channel. He again made use of the *Paramore*, setting out in June 1701 and completing his task in October.

Following this work and at the direct request of Queen Anne (who was the sister of the late Queen Mary and who had been on the throne since 1702 following the death of William III), Halley made a couple of visits to Vienna and the peninsula of Istria, once in 1702–1703 and once later in 1703, where he advised on the fortifications for the two Istrian ports of Trieste and Buccari (now Bakar). When in Vienna, he reported to the Holy Roman Emperor Leopold I, who was apparently so impressed with Halley that he gave him a valuable diamond ring. On his second visit, while passing through Hanover, Halley met and dined with the Prince Elector of Hanover (the future King George I of England*) and his sister, the Queen of Prussia.[21]

Halley was back in England by November 1703, shortly before the "Great Storm" of that month caused much damage and loss of life, forever bringing to an end his travels abroad.

* Some sources say that it was George II, not George I, whom he met, although this seems unlikely.

7

OXFORD

Halley's active life continued throughout the early 1700s. This chapter outlines his contribution to a wide variety of different fields in this period so is necessarily something of a potpourri of unrelated episodes. These range from his abilities both as a mathematician and as a linguist, his original ideas on the deductions one could make from the saltiness of the oceans, his skills at predicting and then portraying a total solar eclipse, his proposal for using Venus to determine the previously uncertain distance to the Sun, and his discovery that the "fixed" stars are not as fixed as was thought.

One further episode, the tragic story of Admiral Sir Cloudesley Shovell and his fleet, is an inevitable part of this chapter given Halley's prescient prediction of a disaster such as this and his role in the attempt to prevent a recurrence.

There were also the inevitable scraps with Flamsteed. Nor would this chapter be complete without a passing mention of one of the biggest scientific rows of all time between Newton and Leibniz in which Halley played a bit part.

PROFESSOR HALLEY (1704)

Halley's failure to obtain the professor of astronomy post at Oxford University in 1691 had been a huge disappointment to him. However, another opportunity came along in 1704. The post of professor of geometry at the university fell vacant following the death of the incumbent, Dr. John Wallis (a mentor, friend, and frequent correspondent of Halley), in late 1703. Halley was now at the height of his career and had just carried out important government work with his two missions to Istria, thereby acquiring even more friends in high places. As the obvious candidate, he was elected to the post in January 1704, this time without the degree of opposition that he had encountered just over 12 years earlier. Both Tillotson and Stillingfleet were by then dead.

Halley would now be spending at least some of his time in Oxford, where he lived in the building at 7/8 New College Lane, shown in figure 7.1. He gave his inaugural lecture in May 1704. A few years after his appointment, in 1710, he would also be given an honorary doctorate from the university. (His correct title therefore became *Dr.* Halley. Contrary to some accounts, he was never *Sir* Edmond Halley.) His years while professor of geometry were among his most fruitful and productive. However, he had to balance his responsibilities here with his election to the Council of the Royal Society the previous November. At this election, he had received four more votes than Isaac Newton, although Newton went on to be elected to the post of president, a post he then held for the rest of his long life.

There was, of course, a disparaging comment made beforehand by Flamsteed about the likely appointment:

> Dr Wallis is dead—Mr Halley expects his place—who now talks, swears and drinks brandy like a sea captain.[1]

But an observation like this was only to be expected from him.

Figure 7.1. Halley's Oxford home at 7/8 New College Lane. The small observatory that Halley constructed on the roof can still be seen. ***Photo courtesy of the author***

APOLLONIUS (1706/1710)

The ancient Greeks were very good at a remarkable number of things, among them mathematics. The ancient Greek mathematician Apollonius of Perga,* who was born in the third century BCE, was one of the most able. One of Halley's duties as the new professor of geometry was to teach the math of ancient Greeks such as Archimedes, Euclid, and Apollonius. An Arabic reconstruction† of Apollonius's work *De Sectione Rationis*, the only version that had survived, had been unearthed in the Bodleian

* Not to be confused with Apollonius of Tyana, who lived in the first century CE and whose birth, life, and death bore a striking resemblance to that of Jesus. The disciples of each teacher argued ferociously with each other about who of the two was really the Son of God and who was merely an imposter. See Ehrman (2008, 20–22).

† It is sometimes referred to as a "translation" and sometimes as a "reconstruction." The latter is probably a more accurate description.

Library at Oxford University by Edward Bernard, the Savilian Professor of Astronomy until 1691. Bernard had tried to translate it into Latin but gave up after completing less than one-tenth of the work, apparently because he found the task too difficult. After his death, Bernard's successor at Oxford (and Halley's friend) David Gregory revised and corrected what Bernard had done and also restored the corrupted Arabic. Then Halley, in spite of not knowing any Arabic at that stage, managed to pick up sufficient knowledge of the language to complete the translation into Latin, a truly remarkable achievement. The translation was published in 1706.[2]

Nor was this Halley's only successful translation. Apollonius had written extensively on conic sections (as discussed in chapter 2). These are slices through a cone that, depending on the angle at which they are made, produce the circle, the ellipse, the parabola, and the hyperbola. He wrote an eight-volume text on this and related subjects simply titled *Conics*.[3] Unfortunately, only the first four books survived in the original Greek, and books 5 to 7 survived only in the Arabic, into which they had been translated many centuries later. Book 8 had been lost. David Gregory and Halley started the work of translating the Greek of the first four books into Latin, but Gregory died in 1708 at the relatively young age of 49. So Halley finished this translation task on his own. In addition, he had acquired sufficient Arabic not only to understand books 5 to 7 but also to translate these into Latin. It is a further mark of his genius that he was able to produce a credible "reconstruction" of the lost book 8 into Latin as well. His efforts were published in 1710.[4]

ADMIRAL SIR CLOUDESLEY SHOVELL (1707)

The Isles of Scilly are a group of small and sparsely inhabited British islands that lie some 45 kilometers (30 miles) to the southwest of the

Cornish mainland. If you pay a visit to their museum, you will learn about one of the greatest disasters ever to befall the British navy* and see many of the objects salvaged from the wrecks of the four ships that sank with massive loss of life. Yet if only those responsible had listened to the advice that Halley had given just a few years previously, the disaster would never have happened.

It was on the night of October 22, 1707,† that Admiral Sir Cloudesley Shovell and a fleet of 22 British ships under his command were journeying home after a military campaign in the Mediterranean. They had sailed from Gibraltar at the end of September, and—given the prevailing winds—traveled roughly northwest into the Atlantic, well to the west of the Bay of Biscay, finally turning eastward on October 18. They were now intending to sail up the English Channel and home to Portsmouth.

However, the sky throughout the day of October 22 had been covered in thick and unbroken rain clouds, accompanied by persistent drizzle. There was a strong wind, and the seas were rough. There was no possibility of observing the Sun to establish latitude. Moreover, the charts they had of the area were inaccurate; most charts of the time showed the Scilly Isles between 22 and 32 kilometers (14 and 20 miles) too far north. Even the most recent chart available showed the Scilly Isles 14 kilometers (nine miles) farther north than the reality. There was probably also a problem with many of the compasses on board the ships. A subsequent inspection of the Admiralty stores found that only 70 out of 370 compasses were of any use, the main problem being rust. The fleet may also have been affected by the Rennell Current, an unpredictable and strong current that could sweep a ship unexpectedly northward.[5]

So the facts of the matter were that the fleet had a poor understanding of both its latitude and its longitude, it was working with dangerously inaccurate charts, its compasses were probably providing inaccurate in-

* The Act of Union between England and Scotland took effect on May 1, 1707. From then on, the English navy became the British navy.

† Britain was well behind mainland Europe in still using old-style Julian dates.

formation, the weather was very poor, and it was dark. The combination could hardly have been any worse and was to prove fatal.

The ships discovered—too late—that, instead of sailing in open waters, they were far too close to the Gilstone Rocks about five kilometers (three miles) off the island of St. Agnes, the most southerly of the larger islands making up the Scillies. The flagship, HMS *Association*, hit the rocks and sank with the loss of the entire crew of some 800 men, including the admiral. A further three ships, the *Eagle*, the *Romney*, and the *Firebrand*, also sank with only one survivor among them: George Lawrence, quartermaster on the *Romney*, was found the following day clinging to a nearby rock. Accounts differ as to how many men were killed in total, but the figure was between 1,300 and 2,000.

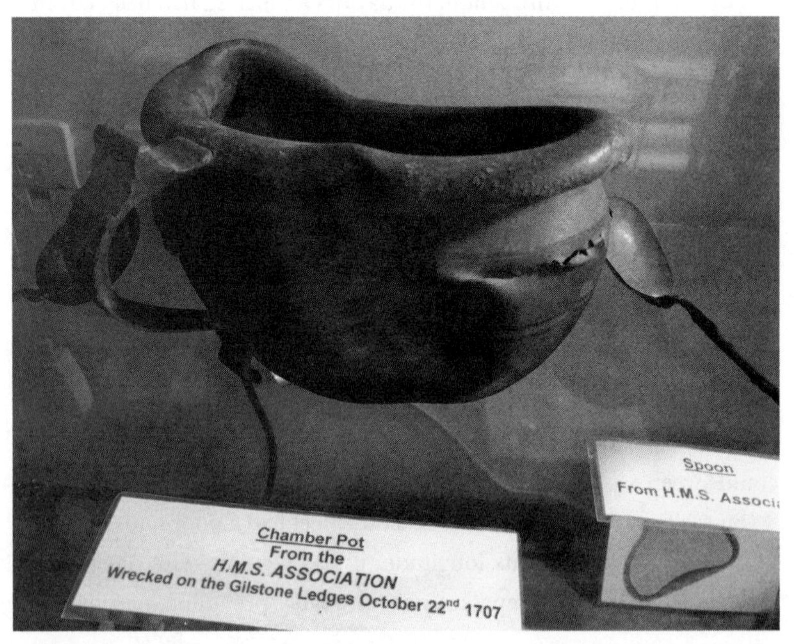

Figure 7.2. A chamber pot and a spoon recovered from the wreckage of the *Association*. *Sandra Greaves/Scilly Isles Museum*

The body of Cloudesley Shovell, along with those of other senior officers, was found (mysteriously) on Porth Hellick beach on the eastern side of St. Mary's island, about 10 kilometers (seven miles) from the site of the shipwreck and in the opposite direction from the prevailing currents. The only credible (but unverifiable) explanation seems to be that he and some officers had escaped from the *Association* in a small boat that had taken them some considerable distance—but ultimately not to safety—and that they had all drowned.

Contrary to one legend, it is not now believed that the admiral had been murdered for his ring, which was missing from his finger. The doctor who carried out a postmortem on his body stated, "Not a mark upon him except for a scratch over one eye as if made by a pin."[6] The ring was, however, never recovered in spite of the reward offered by his widow.

Figure 7.3. A ship's wheel, believed to come from the *Firebrand. Sandra Greaves/Scilly Isles Museum*

So it remains plausible that his ring was stolen from his body, perhaps by one Sally Thomas, a local woman who was probably the first to find the body. The much later story that she had not only stolen the ring but also murdered Cloudesley Shovell and had confessed to this crime on her deathbed is not now generally accepted. Cloudesley Shovell's body was later transported to London and buried in Westminster Abbey (not far from the memorial to Halley, which is in the south cloister) at the personal expense of Queen Anne.

Halley had presciently warned of the danger of such a calamity only a few years earlier. In a paper to the Royal Society,[7] based on his recent experiences of sailing past the Scillies, he had correctly pointed out that the average latitude of the Scillies was 49° 55' north and that their most southerly latitude was 49° 50'.* Unfortunately, in most charts, they were wrongly shown as north of 50°. There was also, he said, a significant magnetic variation (discussed in chapter 6) in this area that at the time was about 7½°. At the time of the disaster, the sailors were unable to check the direction of north in any way other than by a compass. Even if their compasses had been working properly, which most of them probably weren't, if they were unaware of the magnetic variation, they would inevitably have set a slightly erroneous course. Halley had recommended that, for the sake of safety, ships in the area and bound for the English Channel should not travel any farther north than a latitude of 49° 40'. Cloudesley Shovell had failed to follow this advice (if ever he knew about it) and had paid a heavy price.

Contrary to what is asserted in Dava Sobel's *Longitude*, on this occasion an accurate assessment of longitude had been only one of the British fleet's problems. Nevertheless, the disaster was probably one of the main

* Halley's log entry of August 26, 1700, when he was returning to England at the end of his second voyage around the Atlantic and passing by the Scilly Isles, states, "The south part of Scilly lies in 49° 50' *past dispute*" (emphasis added). Halley is clearly making the point that others had incorrectly given a higher latitude but that he—Halley—had carefully checked this measurement and was in no doubt about it.

reasons behind the establishment in 1714 of the Board of Longitude, following a petition to Parliament. Its purpose was to establish an accurate method of determining longitude at sea. It had about a dozen members, one of whom was Halley.

Latitude was relatively easy to measure at sea. Longitude, as discussed in chapters 1 and 6, was an entirely different matter. For an accurate determination of longitude, what was needed was a local clock combined with knowledge of the local time at Greenwich. The difference in the two times effectively gave the longitude. As discussed in chapter 6, there were only two ways of establishing this difference: either a ship had to carry a clock giving the local time at Greenwich, or it had to make use of a "clock in the sky" by, for example, noting the Moon's position relative to background stars. Knowing what that position would be at Greenwich again enabled longitude to be calculated. Halley had been confident that only when the Moon's orbit was accurately known would it be possible to find longitude at sea. At a May 1688 meeting of the Royal Society, for example, he had said that "it would be scarce possible ever to find the longitude at sea, sufficient for sea uses, till such time as the Lunar Theory be fully perfected."[8]

When the Board of Longitude was formed, neither of these methods worked at all well. The best pendulum clocks that ran to time on land quickly went wrong when on a moving ship. And the Moon's precise orbit was so complicated that its position couldn't at that time be predicted exactly in spite of *Principia* providing the theoretical means to do this. (Newton once told Halley that the study of the Moon's orbit was the one thing that gave him a headache.) The Board of Longitude itself, somewhat surprisingly, given the urgency of the problem, didn't have its first meeting until 1737, as will be discussed in the next chapter.

LEIBNIZ (1711/1712)

Isaac Newton was notoriously sensitive to any criticism or questioning, however mild. He was equally sensitive when it came to asserting the priority of his discoveries, even in cases where it was perfectly possible that somebody else had got there first. This was the case for the invention of calculus, which was claimed both by him and by Gottfried Leibniz (1646–1716), a brilliant German mathematician and philosopher and also a Fellow of the Royal Society.

Calculus is an extremely useful mathematical tool. One part of it, differentiation, allows you to find the gradient of a curve at any point. The other, integration, allows you to find the area under a curve and between any two points on it. The modern consensus is that Leibniz and Newton invented calculus independently of each other and at very roughly the same time (something that can often happen in science). In doing so, both men were building on the work of predecessors such as Johannes Kepler.

Newton, unfortunately, didn't believe in giving credit to anybody else where his own inventions and discoveries were concerned. Leibniz had published his ideas on calculus in 1683 in a German journal before Newton published, although Newton's own ideas predated his publication by some years. But—as the recently elected president of the Royal Society—Newton was in a powerful position to set up a committee to establish who had priority in the discovery and to ensure that the committee reached the conclusion that he wanted to see. Not surprisingly, the committee did indeed conclude that Newton had discovered calculus first. Halley was one of the members of the committee and was arguably the most senior on it, so he must have taken a leading role in its findings. The final report was in Halley's handwriting.

FLAMSTEED (1712)

A full-scale row took place in 1712 between Flamsteed and Newton, with Halley once again acting as Newton's coconspirator. The problem had been brewing for several years. Flamsteed was a very capable astronomer but was a perfectionist and had been withholding the observational data, eventually amounting to about 3,000 stellar positions that he had been collecting at Greenwich until he deemed that it was ready for publication. This was in spite of the fact that, for the previous eight years, he had been repeatedly urged by Newton and others to make them public as soon as possible. From his point of view, he was not being unreasonable. After all, he had had to pay for all his equipment and instruments himself, so the results he obtained from them were in a sense his own property.

This was not how everyone else saw the matter. Flamsteed was, they argued, a public servant who was paid his salary out of the public purse, and he therefore had a duty to make his results public as soon as he possibly could. A previous attempt at publication in 1704 had not been successful. So at the instigation of Queen Anne, a Board of Visitors was set up by the Royal Society in 1710 with the aim of visiting and inspecting the observatory on a regular basis and encouraging Flamsteed down the path of publication. Eventually, Queen Anne issued an order for the observations to be published.

At this point, things get a little murky. Flamsteed had made no attempt to get the 1704 results into a fit state for publication, so Halley was asked to edit them as well as he could in preparation for publication. Newton and Halley between them then arranged for some 400 copies of the observations to be printed under the title *Historia Coelestis*. Unfortunately, the publication contained some errors.

Flamsteed was furious. He managed to get hold of some 300 copies of the observations, and publicly burned them* "as a sacrifice to heavenly truth" while continuing to work on his own version. But the genie was out of the bottle.

SALTINESS OF THE OCEANS (1714)

In 1713, Halley became secretary of the Royal Society, succeeding Sir Hans Sloane. Notwithstanding the extra work that this must have produced, together with the requirements of his Oxford post, he continued to churn out a host of significant scientific papers. In 1714, he came up with an ingenious method for establishing the age of the Earth—or at least the age of the oceans. In a paper to the Royal Society, he pointed out that the water in all the lakes in the world gradually evaporates but is replenished by the rivers and streams that run into them. He then noted that those lakes with no means of exit for the water are always salty, whereas those with rivers or streams running out of them are not salty. He concluded that the salt in the lakes must come from tiny amounts of salt in the incoming rivers and streams and that over huge periods of time, these lakes become ever more salty as a result of the evaporation of water from the lakes. The salt can't evaporate, so it remains in the lake.

He then pointed out that the oceans can be thought of as enormous lakes with inputs from all the rivers of the world but with no means of escape for the water other than by evaporation. He stated that the oceans probably started free of salt. So if it were possible to measure the rate at which salt was accumulating in the oceans, by taking measurements of the amount of salt per unit volume at least several hundred years apart,

* He did retain those portions of the 300 copies that he felt were accurate and could be used later for his own publication.

it would be possible to extrapolate backward to determine how long it was since they were salt free. This would then give the age of the oceans.

He conceded that since neither the ancient Greeks nor the Romans had thought of making this measurement for us, his method would work only if measurements were made both in his time and several hundred years into the future. However, as the rate of increase of salt in the oceans was clearly extremely slow and as the oceans were already very salty, he was fully prepared to accept that this would probably result in an age of the Earth that was considerably greater than that derived from the Bible of only about 6,000 years. Rather than concluding that the Bible got it wrong, which would have got him into trouble, he circumvented this difficulty by reinterpreting the length of the first days of creation in Genesis to mean "long periods of time" rather than days of 24 hours that the author of the Genesis story clearly meant. He also concluded that even if the oceans did start with some salt content, the measurement would still give an upper limit to the age of the Earth, thereby disproving the idea that the Earth (or, strictly speaking, the oceans) had existed forever.[9]

A TOTAL SOLAR ECLIPSE (1715)

Total eclipses of the Sun are beautiful sights that do not occur nearly often enough to keep astronomers happy. They happen when the new Moon passes between the Earth and the Sun and the shadow of the Moon falls onto the surface of the Earth. This occurs only when the Moon passes *exactly* between the Sun and the Earth and then only along a narrow path of totality. Total solar eclipses are therefore, to astronomers, infuriatingly infrequent. Although a total solar eclipse happens somewhere about every 18 months, any one spot on the Earth will wait, on average, roughly 375 years between eclipses.[10]

So the people of London were fortunate that on April 22, 1715,* they were treated to one such total solar eclipse, often now referred to as Halley's eclipse, which was observed by Halley and several others on the roof of the recently purchased Royal Society building on Crane Court off Fleet Street in central London.

Halley had used his own lunar observations to calculate in advance, with considerable accuracy, both the path of the eclipse and the time when it would occur. He produced the map shown in figure 7.4, indicating where the shadow would fall. He was not the first to produce an eclipse map, but he seems to have been the first to produce it in this form, with the path of the eclipse seen from above.[11] This is essentially the same as the modern equivalents produced by the eclipse expert Fred Espenak.[12]

He circulated the map widely in advance of the eclipse. This was partly to allay fears among the general populace that any harm would come to them or to the new Hanoverian king, George I (who had come to the throne the year before following the death of Queen Anne), as a result of the eclipse. It was, he said, an entirely natural and predictable event.† He afterward collected and published a huge number of observations of the eclipse from all over the country and from abroad.[13]

Needless to say, Flamsteed made a caustic comment on a minor inaccuracy in Halley's map, although Halley did correct this later.[14]

While any one spot on the Earth has to wait on average about 375 years between total solar eclipses, this is only an average. A further total solar eclipse whose path also crossed part of England took place only

* This is the old-style Julian date still in use in England. The date of the eclipse is sometimes quoted as May 3, 1715, which is the new-style Gregorian date used in continental Europe but not adopted in Britain until 1752.

† This is the quotation from the first paragraph of Halley's words under his map of the eclipse: "The like Eclipse having not for many ages been seen in the Southern Parts of Great Britain, I thought it not improper to give the Publick an Account thereof, that the sudden darkness, wherein the Starrs will be visible about the Sun, may give no surprize to the People, who would, if unadvertised [i.e., if not warned], be apt to look upon it as Ominous, and to interpret it as portending evill to our Sovereign Lord King George and his Government, which God preserve. Hereby they will see that there is nothing in it more than Natural, and no more than the necessary result of the Motions of the Sun and Moon; And how well those are understood will appear by this Eclipse."

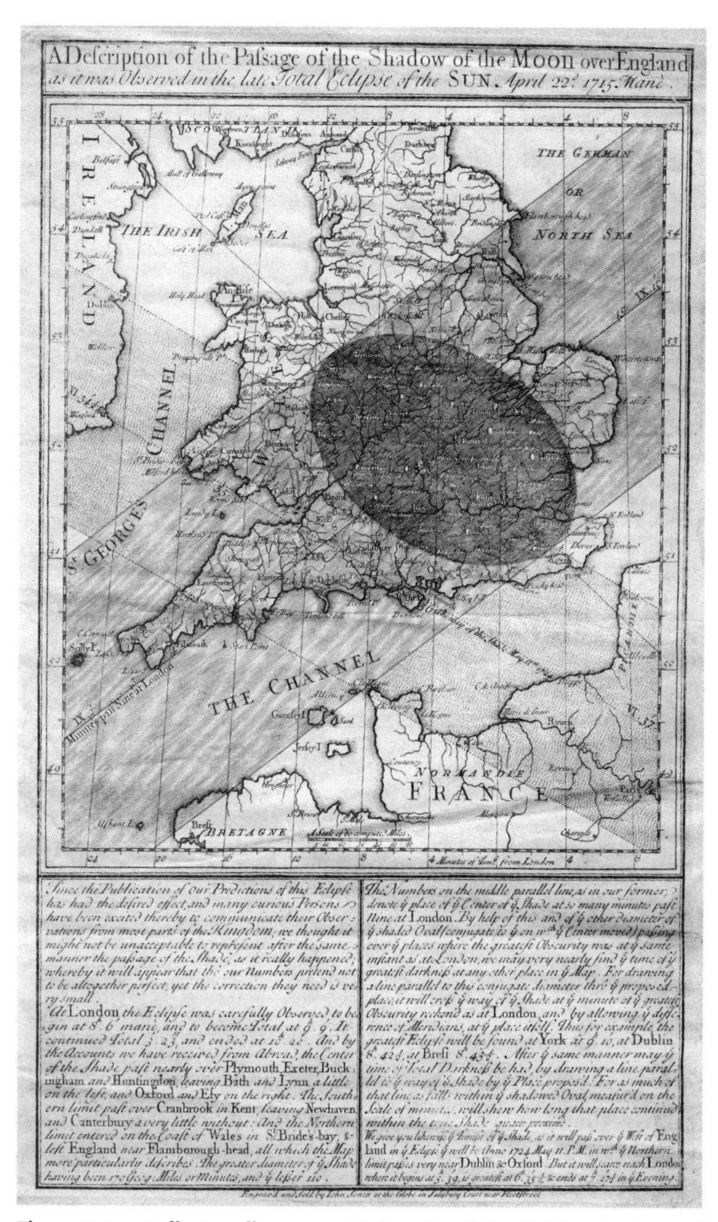

Figure 7.4. Halley's eclipse map. ***University of Cambridge, Institute of Astronomy Library***

about nine years later on May 11, 1724, and Halley once again produced a map of it in advance. (The path of this later eclipse is also shown in figure 7.4. The two paths cross each other, almost at right angles, over a large part of southern England and Wales. People living in these areas were fortunate in being able to see two total solar eclipses separated by only nine years, a rare combination of events.)

VENUS (1716)

In 1716, Halley presented a paper to the Royal Society, published in *Philosophical Transactions*, showing how a transit of the planet Venus across the face of the Sun could be used to measure the distance to the Sun, something that was previously poorly known.

Around 1672, both Cassini and Flamsteed had independently used observations of Mars to derive a figure for the Earth–Sun distance that was much greater than had previously been assumed and turns out to have been only about 8 percent lower than the modern value. But it was difficult to make these observations of Mars, so it was not possible to be certain of their results.

Ever since he had observed the transit of Mercury across the face of the Sun in 1677, while he was on St. Helena and possibly before that, he had realized that transits could be used to measure the Earth–Sun distance. His paper of 1716 set out the detailed mathematics of how this could be done. He had published previously on the subject, but this paper was more thorough. The key point that Halley had recognized was that it was (in principle) a lot easier to measure accurately the times of a transit of Venus than to measure accurately the positions of Mars, as Cassini and Flamsteed had done. Different observers, in different places, would observe different transit times, and a complex calculation using these different times would then provide the distance to the Sun. However, what would be required would be scientists all over the planet making observations of the times of the transit at each location.

One problem with this proposed method is that transits of Venus don't happen very often. Venus is, of course, in an orbit closer to the Sun than the orbit of the Earth and takes about 225 days to go once around the Sun. This means it will overtake the Earth every 584 days on average. If Venus and the Earth orbited in *exactly* the same plane, transits would therefore occur every 584 days.

But most of the time, Venus passes either just above or just below the Sun as seen from the Earth. In fact, transits occur in pairs that are only eight years apart, and those pairs are in turn separated by more than 100 years. The last pair of transits was in 2004 and 2012, but the next pair will not be until 2117 and 2125.

Halley realized that an ideal opportunity would come in the pair of transits due in 1761 and 1769. He knew that he would be dead long before these occurred, so he issued a clarion call to scientists all over Europe to take full advantage of the transits:

> I recommend it therefore again and again to those curious astronomers who, when I am dead, will have an opportunity of observing these things, that they remember my admonition, and diligently apply themselves with all imaginable success.

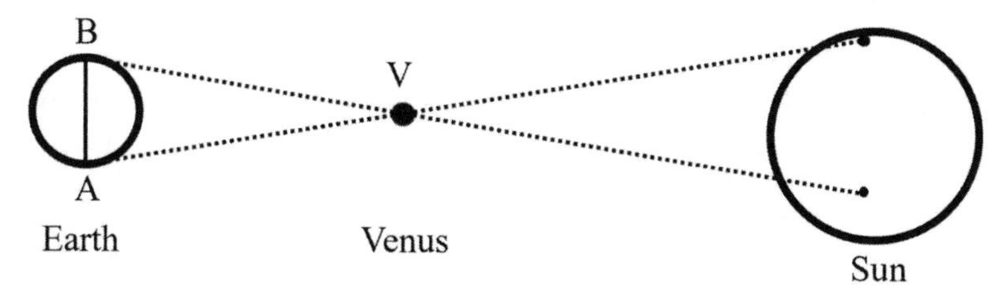

Figure 7.5. Observers at A and B see Venus disappearing off the edge of the Sun at different times (not to scale). ***Created by author***

His 1716 paper was deliberately written in Latin to ensure maximum publicity throughout Europe for the idea. As we shall discover in chapter 9, the international scientific community would rally enthusiastically to his call.[15]

THE FIXED STARS (1718)[16]

For thousands of years, objects in the sky had been thought of in two categories: the "fixed stars" that provided a fixed and unchanging background in the sky and the seven objects that constantly moved against this background: the Sun, the Moon, Mercury, Venus, Mars, Jupiter, and Saturn. (Uranus would not be discovered until 1781 and Neptune not until 1846.)

Until just before the time of Halley, the fixed stars had been thought to be attached to a very distant celestial sphere centered on and rotating around the Earth about once every 24 hours. After Copernicus had put forward the idea of a rotating Earth that moved around the Sun, the idea of a celestial sphere lost its raison d'être and gradually lost plausibility. Instead, it was replaced by the view that the stars were independent objects that nevertheless remained fixed in space in their relative positions.

Halley was the person who discovered that this was wrong.

In 1718, he published a paper for the Royal Society in which he announced that he had examined the positions of several stars that had been cataloged by the ancient Greek astronomers Hipparchus (ca. 190–120 BCE) and Ptolemy (ca. 90–170 CE) and discovered that, over the intervening 1,800 or so years, three of the brightest stars in the sky had changed their positions. These were Aldebaran in the constellation of Taurus, Sirius in Canis Major, and Arcturus in Boötes. All three stars, he said, were roughly half a degree (roughly the apparent diameter of the full Moon) farther south than the ancient Greeks had recorded. He correctly recognized that their brightness compared with most other stars indicated that they were probably relatively close by. So the most likely explanation for their observed movement was simply that this would be easier to detect than the motion of more distant stars.

This was an important discovery and a revolution in our understanding that paved the way for the much later recognition that we live in a far from static galaxy of stars, all of which rotate round the galactic center but also have a degree of random motion. It was a final confirmation that the stars were not in fact attached to a static celestial sphere, as had been believed since antiquity and had still been accepted, for example, by Kepler.

Halley's figures line up tolerably well with modern measurements for the movements of Sirius and Arcturus but not for Aldebaran. So in retrospect, it seems that Halley may have made an error in his calculations for Aldebaran.[17] However, although it has been disputed,[18] his discovery of the motions of Sirius and Arcturus represented a major step forward in our understanding.

Halley made this discovery while investigating the "precession of the equinoxes." The Earth's axis does not constantly point at one spot in space. Instead, driven by the effects of the Sun's and Moon's gravity on the Earth's equatorial bulge, the axis "wobbles" slowly over a period of some 26,000 years, rather like the wobbling motion of a child's toy top. This phenomenon had been known since antiquity thanks to the work of the ancient Greek astronomer Hipparchus, but had never been explained until Newton's *Principia* did so.

Although Halley would continue in his post at Oxford for the rest of his life, from 1720 on, his workload elsewhere substantially increased because of the (probably inevitable) appointment that came his way in that year, the subject of the next chapter.

8

GREENWICH

Flamsteed would have been livid. Yet he probably knew it was inevitable. He died on December 31, 1719, at the age of 73 and after nearly 45 years as the first Astronomer Royal. And Halley, clearly by far the most suitable person to replace him, was appointed in his place early in 1720 at the advanced age of 63.

Flamsteed did get in a bit of posthumous revenge. His widow, acting on instructions he gave her before he died, ensured that all the instruments were removed from the observatory. This was not entirely unreasonable, as all these instruments had been purchased at Flamsteed's expense and were therefore his property. As Halley said of his arrival, he found the observatory "wholly unprovided of instruments, and indeed of everything that was moveable." A more generous individual than Flamsteed might have decided to leave the instruments in situ for the benefit of the nation. Instead, Halley had to buy new ones. His other step, taken shortly after his arrival at Greenwich and to enable him to concentrate on his work there, was to resign from his post as secretary of the Royal Society.

It needs to be made clear that Flamsteed did leave one important item to posterity: his catalog of nearly 3,000 accurately measured stellar positions, the three volumes of which were finally published posthumously in 1725 under the title *Historia Coelestis Britannicae* by his devoted assistants Abraham Sharp and Joseph Crosthwait. There is no doubt that, in spite of his faults, he was a careful and assiduous observer. The catalog was an extremely valuable contribution both to science and to navigation and was the standard star catalog for astronomers for the next century.

OLBERS' PARADOX: WHY IS THE SKY DARK AT NIGHT?

Only a few months after his appointment as Astronomer Royal, Halley endeavored, in two papers to the Royal Society,[1] to tackle the problem that subsequently became known as Olbers' paradox: why is the sky dark at night?

The answer to this question might at first seem blindingly obvious: because the Sun isn't shining. Unfortunately, it isn't that simple, and attempts to find the correct answer have a long history, of which Halley's contribution was only one part. The following paragraphs explain the paradox and outline Halley's contribution to a debate that has spanned the centuries.

The hundreds of billions of stars in our Milky Way Galaxy are arranged in a spiral-shaped disk. However, for the purpose of explaining the paradox, we can think of them as being more or less randomly distributed in space. Beyond our Galaxy are hundreds of billions of other galaxies, also randomly distributed. To understand the paradox, the stars and the galaxies beyond them can be imagined as occupying a series of concentric shells centered on the Earth, as shown in figure 8.1. If you do the math, you discover that, perhaps surprisingly, the amount of light re-

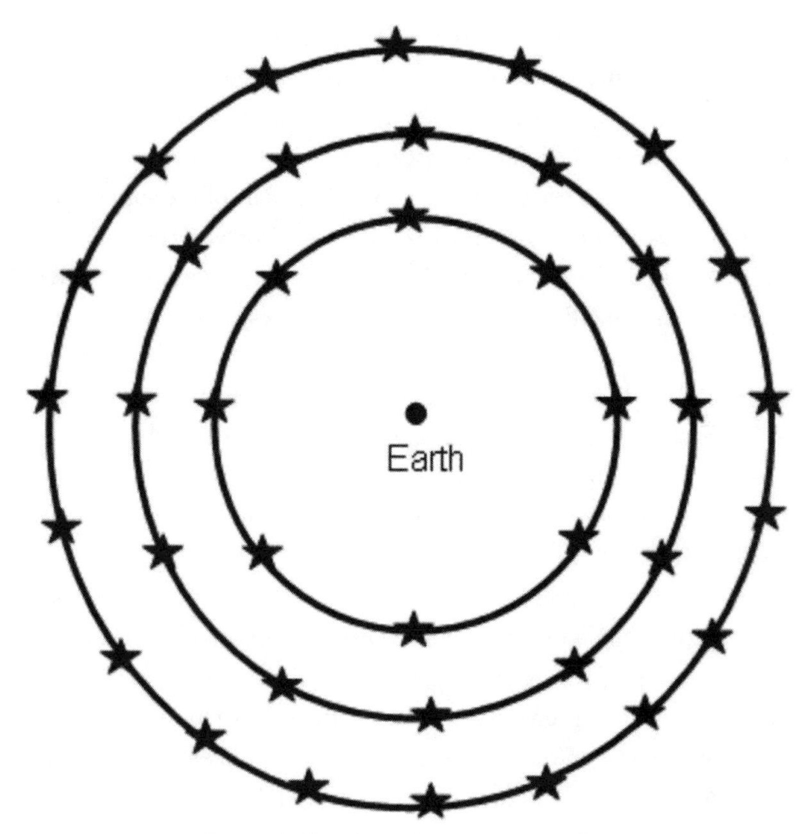

Figure 8.1. To understand Olbers' paradox, we can treat the stars and the galaxies beyond them as though they formed a series of concentric shells around the Earth. ***Created by author***

ceived by the Earth from each shell, no matter how distant, is the same as that received from every other shell, no matter how close. This is because the increasing distance of each shell is exactly balanced out by the extra light generated in larger and larger shells. More and more shells will cause more and more light to reach the Earth.

So the problem is that if the Universe is infinite in extent and has existed forever, the amount of light from all these shells should be infinitely great. Far from being dark at night, the sky should be blazing with light.

But it isn't. The one obvious fact about the night sky is that it is dark. So it cannot be the case that the Universe has existed forever and is infinite in extent. It is a remarkable fact that such a profound conclusion can be deduced from such a simple observation.

What is the best resolution of the paradox? Part of the answer must lie in the fact that the Universe has not existed forever, certainly in its current form, but has a finite age, 13.8 *billion* years, according to current measurements, and that it is expanding, which means that there is a limit to how far we can see.

One of the first people to tackle the problem was Johannes Kepler, who proposed that we live in a finite Universe in which there is a gigantic celestial sphere that is centered on the Sun and that provides the Universe's outermost boundary.[2] By Halley's time, this was no longer credible.

Halley's first contribution was to point out that the Universe certainly seemed to be infinite because more powerful telescopes revealed more and more stars, presumably at greater and greater distances. Furthermore, he saw that a problem would arise if there were only a single group of a finite number of stars in the Universe. The problem would be that gravitational attraction between these stars would cause them to fall inward toward their common center and eventually to coalesce into a single object. He realized that this implied that the Universe must be infinite in extent (a view also held by Isaac Newton). The lack of an edge beyond which stars did not exist would prevent any such collapse.

Halley recognized that this created another problem. If the Universe of stars extended indefinitely, which he thought must be the case, then this would potentially lead to an infinitely bright night sky, which clearly wasn't the case. He saw that this was therefore an argument against the idea that the Universe has existed forever. He went on to argue (incorrectly) that if the Universe were infinite in extent, the light from more distant stars would not mean ever-increasing levels of light arriving at the Earth because these stars were individually so small.

Since Halley's time, others who have tackled the problem include Heinrich Olbers (1758–1840), the German astronomer whose name has been given to the paradox; the scientist Lord Kelvin (1824–1907); the cosmologists Hermann Bondi (1919–2005) and Edward Harrison[3] (1919–2007); and even the American author Edgar Allan Poe (1809–1849).

RECOGNITION (1729)

Throughout his life, Halley never seems to have had any financial worries. However, his new salary of £100 a year as Astronomer Royal was supplemented as a result of the visit of Queen Caroline, wife of King George II, to the Greenwich Observatory in 1729. (George I had died in 1727.) She was impressed by her visit and noted that Halley had previously (albeit for only a couple of years) been a captain in the Royal Navy, so she persuaded George to grant Halley a pension equal to half the pay he had received as captain.[4] This was not the only recognition to come his way; also in 1729, he was admitted to the French Académie des Sciences in Paris as a foreign member.

LONGITUDE: LUNAR DISTANCE METHOD (1731)

The problem at that time of how to find longitude at sea has been a recurring theme in this book. Halley's view had always been that the best way of discovering this was by observing the position of the Moon relative to the background stars, the so-called lunar distance method. When he had begun his career more than 40 years previously, neither the positions of the fixed stars nor the orbit of the Moon were sufficiently accurately known. While he was Astronomer Royal, Flamsteed had successfully carried out the first part of this work but had not done much by way of establishing a definitive orbit of the Moon.

Halley's major program in this period, therefore, was to chart as accurately as possible the Moon's 18-year cycle. This cycle, known as the Saros period, needs an explanation. It might very reasonably be thought that the Moon simply repeats its orbit around the Earth once every (lunar) month of about 29½ days. In fact, the complicated geometry of the Earth–Sun–Moon system means that it doesn't truly repeat itself until after a period of 223 lunar months (or 18 years and 11.3 days) has gone by. At the end of this period, the Moon's orbit does finally repeat. So, for example, a particular sequence of solar and lunar eclipses will then repeat.

At age 63, Halley thought that he would not live long enough to observe the Moon for the next 18 years. In fact, he did, with a few years to spare. After nine years, at the halfway point in the cycle in 1731, he presented his results to the Royal Society.

In his paper,[5] he said that nearly 50 years previously, just after his marriage and his move to Islington, he had begun the task of observing the Moon's orbit more precisely, because of the near universal agreement that this would provide the only way to discover longitude at sea. At that time, he had accumulated nearly 200 observations and from these had made the interesting discovery that the irregularities in the Moon's orbit occurred regularly. In other words, there was a clear pattern to the irregularities. This meant that his observations had been sufficient for him to be able to predict the degree of error in the standard lunar tables of the time. Unfortunately, he said, family matters had intruded and prevented him from completing a full cycle of observations. (He was referring to the death of his father in mysterious circumstances in 1684 and the quarrels with his stepmother about their respective entitlements to his father's property.)

This time, by 1731, after making about 1,500 observations of the Moon relative to the stellar background over the previous nine years, Halley was able to provide predictions of the Moon's position that would enable a measurement of longitude at sea to within 20 leagues (about 100 kilometers, or 60 miles, depending on the precise definition of a league) at the equator and to less than 15 leagues as far north as the English Channel. Unimpressive as this sounds to modern ears, it was still a significant improvement on what had gone before and seemed to bring the problem of the determination of longitude at sea within sight of a solution.

Halley's aim was to use his observations to construct improved lunar tables. However, in spite of criticism from Newton, he insisted on delaying publication of these until he felt they were ready. The tables ended up being finally published only in 1749, seven years after his death, by his friend John Bevis.

LONGITUDE: CHRONOMETERS

Halley was not wedded to the lunar method to the complete exclusion of all other possibilities. In 1728 at the Greenwich Observatory, he had received a young guest by the name of John Harrison, a Lincolnshire carpenter and clockmaker with little by way of formal education. Harrison had resurrected the idea of measuring longitude by constructing a chronometer that would be set to the time at Greenwich and would be capable of withstanding a sea voyage without gaining or losing time. He even had provisional designs for how such a chronometer would work. The difference between the local time at sea and the time shown on the chronometer would immediately give a measure of the ship's longitude.

Halley, as was his usual manner, received Harrison politely and attentively. However, he cautioned against calling for an immediate meeting with the Board of Longitude, which was likely to be skeptical of anything other than an astronomical method for finding longitude. Instead, he referred Harrison to an instrument maker and fellow Royal Society member by the name of George Graham.

This, it turned out, was the best thing that Halley could have done. George Graham gave John Harrison huge encouragement and a very helpful loan. After much work, Harrison was able to present the first of his chronometers to the very first meeting of the Board of Longitude (of which Halley was also a member) in 1737. The story of Harrison's further endeavors and his eventual triumph is fairly well known but will not be repeated here, as Halley was dead long before Harrison's final victory.[6] However, from all that we know about Halley, we can safely say that, unlike Nevil Maskelyne, one of his successors as Astronomer Royal, he would have welcomed Harrison's success. Nor would he have been in the least upset that the lunar method on which he had worked so hard would eventually become redundant as a result of Harrison's work.

THE ROYAL SOCIETY CLUB (1731)

Halley had always been a highly sociable individual who delighted in the company of others, so although the precise origins of the Royal Society Club (a dining and conversation club for Royal Society members) are uncertain, it appears very likely that Halley had something to do with it, and he has often been described as its founder. He had been for many years, even before his appointment as Astronomer Royal,[7] a regular visitor to Child's Coffee House in central London and close to St. Paul's Cathedral. He and a few others subsequently moved their meetings to a house on Dean's Court, also close to St. Paul's and not far from the Royal Society's headquarters at the time on Crane Court off Fleet Street. Here, they had regular meals together. We are told that Halley, by 1731, at the age of around 74, never ate anything except fish because he had lost all his teeth. They later met nearby at the King's Arms. From what limited information exists, this seems to have been the origin of the Royal Society Club, which may (or possibly may not) have been formally established in 1731 and which still has regular meetings even today.[8]

HALLEY'S DEATH

Halley's good friend Isaac Newton had died in 1727 at the age of 84 following a few years of declining health, one feature of which was the embarrassing problem of urinary incontinence. In the last few weeks of his life and after decades of genuine friendship, there was an unfortunate public quarrel between the two because of Halley's reluctance to publish his lunar observations before they were ready. It is ironic that, in his refusal to publish, Halley was doing exactly the same thing for which he and Newton had criticized Flamsteed.

Although deeply religious, Newton did not accept some of the main tenets of Christian belief and refused to receive the sacrament of the church when on his deathbed. His body lay in state in Westminster Abbey, where his tomb now is.

In the last few years of Halley's life, in 1736, his wife Mary died and was buried in the cemetery of St. Margaret's Church in the small village of Lee (now part of southeast London) not far from Greenwich. In 1741, his son Edmond (who had been a surgeon in the Royal Navy) also died, at the early age of around 43. Halley himself, after a lifetime of remarkably good health, developed a paralysis of the right hand in his early eighties. This did not completely bring his observing schedule to an end, but it did mean that he needed the help of a friend to write down his results:

> He came still as usual once a week, till within a little while before his death, to see his friends in town on Thursday, before the meeting of the Royal Society. This pleasure he continued to enjoy to the last verge almost of his

> life. . . . He was of a happy constitution, and preserved his memory and judgment to the last, as he did also that particular cheerfulness of spirit for which he was remarkable.
>
> His paralytic disorder gradually increasing, and thereby his strength wearing, though gently, yet continually away, he came at length to be wholly supported by such cordials as were ordered by his physician, till being tired with these he asked for a glass of wine, and having drunk it presently expired as he sat in his chair.[9]

Halley died on January 14, 1742. He had lived through the reigns of six monarchs.* In his long life, he had accomplished a huge amount in fields that went well beyond the astronomy in which he had specialized. If it had not been for Isaac Newton's massive contributions, which inevitably overshadowed those of everyone else, he would almost certainly have been seen as the outstanding English scientist of his day.

In the centuries after his death, he has had one crater on the Moon and one on Mars named after him in honor of his contributions to astronomy. The Halley Research Station in Antarctica is also named after him, an indication of his contributions to geophysics. In 1986, to coincide with the return of Halley's Comet, a memorial to him was unveiled in the south cloister of Westminster Abbey. The delay of more than 200 years in setting up this memorial may have been because of the excessively broad religious views that he was believed to have held—but better late than never.

Halley was buried in the cemetery at Lee in the same grave as his wife. The grave can still be visited, although it and the surrounding cemetery are now in a distinctly poor state of upkeep. The memorial that used to cover the tomb was removed in 1854 and placed on prominent display in an upright position in the courtyard at the Royal Greenwich Observatory.

* Or seven or five, depending on how you count them. William and Mary reigned jointly, and Halley died before the reign of George II came to an end. The monarchs were Charles II, James II, William III and Mary, Anne, George I, and George II.

Figure 8.2. Halley's grave, in the cemetery of St. Margaret's Church, Lee. ***Photo courtesy of the author***

Figure 8.3. Halley's tombstone (bottom right of photograph), inset into a wall at Greenwich Observatory. *Royal Museums Greenwich*

Here it acts as a fitting tribute to one of the greatest astronomers of the seventeenth and eighteenth centuries.

Halley's early French biographer Jean-Jacques d'Ortous de Mairan deserves the final word:

> Halley had many qualities that endeared him to others. First of all, he had a genuine affection for them; naturally full of fire, his mind and his heart were animated in their presence with a warmth that the mere pleasure of seeing them seemed to produce. He was frank and decided in his proceedings, equitable in his judgments, equal and settled in his morals, gentle and affable, always ready to communicate. . . . He opened the way to riches for others by all he did to advance navigation, made more remarkable by never having done anything to become personally rich.
>
> He was generous, and his generosity was exercised even at the expense of a vanity from which scholars are no more exempt than other men.
>
> Finally, qualities so rare and so estimable were seasoned in Halley with an essential happiness that neither old age nor the paralysis which he suffered a few years before his death, could ever alter; and this naturally happy disposition was the result of an inner contentment which sprang from an underlying integrity.[10]

9

LEGACY

Venus and Beyond

Halley's mark on science has been considerable, and his contribution to astronomy was only one aspect of his huge influence, as the previous chapters have shown. His two major achievements, as midwife to *Principia* and as the person who demonstrated that comets return again and again, have been dealt with in chapters 2 and 3. But his influence extended well beyond his death in one other important respect—his recommendation to the international scientific community that they should use the upcoming transits of Venus across the face of the Sun to determine the previously rather uncertain distance from the Earth to the Sun.

THE DISTANCE TO THE SUN—AND THE CONSEQUENCES

Measuring distances beyond the Earth is tricky. The distance to the Moon, easily our closest neighbor, can be measured by a method called parallax. Two observers separated by as great a distance on the Earth's surface as possible see the Moon in different positions in the sky relative to the background stars. They can use this information plus the known distance between them plus a bit of math to work out its distance. The Moon's distance has been known reasonably well since it was first calculated by the ancient Greek Hipparchus, who established that the Moon's average distance from the Earth is roughly equal to 30 Earth diameters.

Beyond this, things became considerably more difficult. Within our Solar System, ever since Copernicus and Kepler, relative distances were known but not absolute distances. And the distances to the stars were completely unknown. What was needed was to find out the average distance of the Earth from the Sun. Once this fundamental baseline had been established, it would start to be possible to get a grip on the size of our Universe.

In 1672, the Italian astronomer Giovanni Cassini, whom we met in chapters 1 and 7, used observations of Mars, combined with the known relative distances of Mars and the Earth from the Sun, to calculate a tolerably good Earth-Sun distance of about 137 million kilometers (about 86 million miles), much better than had been obtained before and only about 8 percent below the true figure. So Cassini had shown that the Earth-Sun distance was about 20 times as big as the figure that Copernicus and Kepler had supposed. Flamsteed obtained a similar (but not quite as good) result at about the same time.[1]

Halley thought it would be possible to do significantly better than this by making use of observations of Venus on the rare occasions when it transited across the face of the Sun. An outline of what Halley was proposing was set out in chapter 7. The mathematical details of the calculations are beyond the scope of this book. However, the central point is that if at least two different observers—at different longitudes on the Earth and as widely separated as possible—can note the exact time at their particular location when Venus starts crossing the Sun, then a complex calculation will reveal the distance to the Sun.

The international scientific community responded enthusiastically to Halley's suggestion, and numerous expeditions were mounted to observe both the 1761 and the 1769 transits. Observing sites were set up in several places, including Siberia, California, Hudson Bay, Madras, and Tahiti as well as around Europe.

TAHITI

The 1769 expedition to Tahiti was commanded by Captain James Cook. On board was the astronomer Charles Green, together with several other scientists from different disciplines, to take advantage of the opportunity to discover what they could of the new lands, plants, and animals that they would undoubtedly discover. The scientists were funded by the Royal Society. In addition to Tahiti, they visited several other Pacific islands, Australia, and New Zealand, returning to England in 1771. The ship was provided by the Royal Navy, which also supplied a large contingent of well-armed sailors and marines. Given that much of the data collected had significant military or political value, it is worth asking, as Yuval Noah Harari's book *Sapiens* asks, whether this was "a scientific expedition protected by a military force, or a military expedition with a few scientists tagging along." It was, of course, and, as he says, both.

INTERLUDE: LE GENTIL

No description of the efforts to observe the 1761 and 1769 transits of Venus would be complete without mention of the tragic tale of an astronomer with the lengthy name of Guillaume Joseph Hyacinthe Jean Baptiste le Gentil de la Galaisiré (Guillaume to his friends and Gui to his wife).

He was sent by the French Académie des Sciences to observe the 1761 transit in Pondicherry in India. Unfortunately, Pondicherry had been captured by the English just before his arrival, and he wasn't allowed to land there. Although he was just about able to see the event, the fact of being on the deck of a moving ship meant that he couldn't obtain precise timings of the transit, and his observations were of no scientific use.

Rather than return all the way home and then come back for the next transit eight years later, he decided to stay in the area and wait for 1769. His calculations showed that Manila would be a much better spot for observing this transit. Unfortunately, he was ordered to stay in Pondicherry. The inevitable happened—on the day of the transit, the skies were beautifully clear in Manila but cloudy in Pondicherry. So poor old Guillaume had wasted his time.

Worse was to come. On returning home, after a total absence of well over eight years, he found that he had been presumed to be dead, and his relatives were dividing up his estate between them. Only after a lengthy court battle did he manage to get back most (but not all) of his possessions.

After that, perhaps not surprisingly, he gave up astronomy.

THE RESULTS

The results from the large number of transit observations by the international scientific community were not as clear-cut as had been hoped. It turned out to be difficult to decide on the exact moment when the whole disk of Venus became visible against the Sun's disk. This was because of something known as the "black drop" effect. The atmosphere of Venus causes a black drop to appear between the planet and the Sun's edge, making it difficult to determine the exact time that the transit begins or ends. Nevertheless, they confirmed that the Earth–Sun distance was even greater than Cassini and Flamsteed had calculated. The final result was within some 2 percent of the modern figure of 150 million kilometers (93 million miles).

Halley's proposal had been thoroughly vindicated. Thanks to him, the size of our Solar System was now known better than ever before. Using this and the more accurate figures that became available over time as the base, it would now be possible to measure the distances to the stars and beyond.

THE DISTANCES TO THE STARS

Once the Earth–Sun distance had been established and once the technology had improved sufficiently, it became possible to measure the distance to the nearest stars. This uses the method of parallax to observe the very tiny differences in the position of a nearby star against the background of much more distant stars. As can be seen in figure 9.1, the two positions of the Earth six months apart (say, in January and July) form the base of a triangle with a nearby star. The tiny movement of this nearby star relative to more distant stars behind it enables the calculation of the angles in this triangle. Thanks to Halley, the size of the base of the triangle (twice the Earth–Sun distance) was now known tolerably well, so some simple math enabled the calculation of the distance to the star.

The first result to be accepted by the astronomical community, arrived at by the German astronomer Friedrich Bessel (1784–1846), was published in 1838. He showed that the distance to his chosen star, called 61 Cygni, was about 10.5 light-years, very close to the modern value. (Distances in the Universe are so large that rather than stating them in kilometers or miles, they are usually given in units of the distance that light travels in one year [one light-year]. If stated in kilometers, the

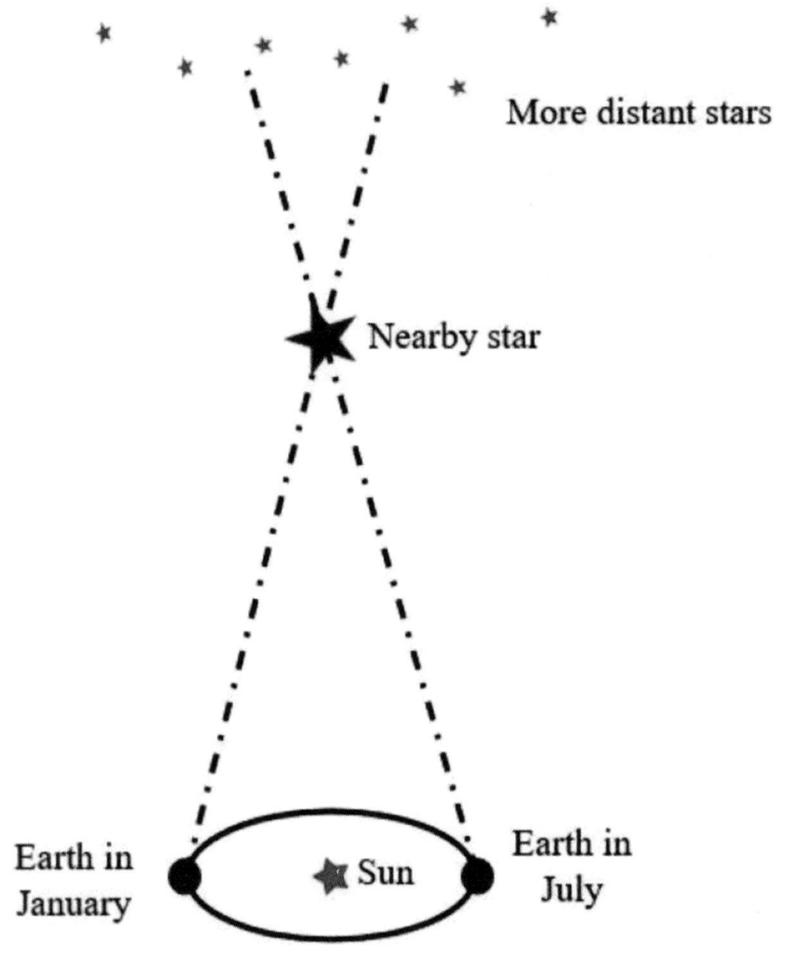

Figure 9.1. How to measure the distance to a nearby star. ***Created by author***

distance to even the nearest star would be the rather unwieldly number of 110,000,000,000,000.)

Today, we have used this parallax method to measure the distances to roughly 1 billion stars, thanks mainly to the observations from *Gaia*, a European Space Agency satellite launched in 2013.

THE DISTANCES TO THE GALAXIES

Until the 1920s, astronomers were puzzled by the nature of the fuzzy white patches that can be seen in the night sky. One such example is the Andromeda Nebula, which can just be seen by the naked eye as a fuzzy white patch in the constellation of Andromeda. (The word "nebula" is simply the Latin word for "cloud.") Another was what we now refer to as the Whirlpool Nebula, or M51, observed in detail by William Parsons (1800–1867), the Third Earl of Rosse, in 1845, with his gigantic telescope at Birr Castle in what is now the Republic of Ireland. Parsons discovered that M51 had a beautiful spiral structure. His sketch of M51 was remarkably close to the images seen in modern photographs (see figures 9.2 and 9.3).

The big question was whether these were objects within our own Milky Way Galaxy (the majority view) or whether they were perhaps distant galaxies in their own right (the minority view). The puzzle became soluble thanks entirely to the work of the American astronomer Henrietta Swan Leavitt (1868–1921). She was studying particular types of stars called Cepheid variables. As the name perhaps indicates, these are stars that oscillate regularly in brightness over periods of days or weeks. In

Figure 9.2. The sketch of M51 by William Parsons. ***Wikimedia Commons***

Figure 9.3. M51, the Whirlpool Nebula. ***NASA, ESA, S. Beckwith (STScI), and the Hubble Heritage Team (STScI/AURA)***

particular, she was studying the Cepheid variables that she had been able to detect in the Large and Small Magellanic Clouds, two clouds of distant stars that we now know are independent galaxies that orbit round our own Milky Way Galaxy and that Halley had observed while on St. Helena. These can be seen only from the Southern Hemisphere. In 1912, she published figures for both the apparent brightness and the periods of oscillation of 25 Cepheid variable stars in the Small Magellanic Cloud.

The point about the apparent brightness of a star is that it varies, depending on how far away the star is. If you compare two stars known to have identical luminosities (i.e., identical intrinsic brightnesses) and one is twice as far away as the other, the farther star will have one-quarter of the apparent brightness of the nearer star. This is because the intensity of light (just like the strength of a gravitational field) diminishes with the square of the distance. So simply determining the apparent brightness of a star as it appears from the Earth gives no necessary indication of its distance, as stars vary considerably in their intrinsic brightnesses as well as their distances.

However, the key point about the Small Magellanic Cloud is that it is so far away that all the stars in the cloud are effectively the same distance from the Earth. In the same way, if you were measuring the distance from New York to Paris, it really wouldn't matter if you measured the distance to the Eiffel Tower or to Notre Dame Cathedral. The difference in distance to these two places is tiny in comparison with the overall New York–Paris distance. So if one Cepheid variable in the Small Magellanic Cloud has, say, twice the apparent brightness of another, its absolute luminosity is also twice as great.

When Leavitt placed her figures for the Cepheids in order of their periods of oscillation, she spotted that their average apparent brightnesses were also in the same order. The longer the period of oscillation, the brighter the Cepheid star. And because all these stars were effectively at the same distance, this meant that there was also a relationship between their periods of oscillation and their absolute luminosities. This was an extremely important discovery. It meant that once you had calibrated the scale, all you had to do was to measure the period of oscillation of a Cepheid variable star, and you would automatically know its absolute luminosity. You can observe its apparent brightness directly, so a simple bit of math then enables you to calculate the distance to the star.

Leavitt certainly deserved a Nobel Prize for this discovery. She would probably have received one were it not for the fact that she died from stomach cancer at the tragically young age of 53, and Nobel Prizes can be given only to people who are still alive. But her discovery was the key that enabled Edwin Hubble (1889–1953), another American astronomer, to unlock the fact that many nebulae were in fact enormously distant galaxies in their own right. He did this simply by observing Cepheid variables in these galaxies and doing the necessary math. For example, we now know that M51 is about 25 *million* light-years away.

By measuring the speeds with which several galaxies were moving away from our own galaxy and comparing this with their distances, Hubble was also able to establish that our Universe is expanding.

We now know that our Universe began in a Big Bang about 13.8 *billion* years ago and has been expanding ever since. It is vastly larger than what was believed only 100 years ago. If the theory of the multiverse is correct, it may also be just one among billions of other universes. A detailed account of how these conclusions have been arrived at can be found in a number of the books listed in the bibliography. However, it is hoped that this book will have given readers some idea of the important role that Halley played in helping to bring about our modern understanding.

APPENDIX A

Edmond Halley's Family Tree

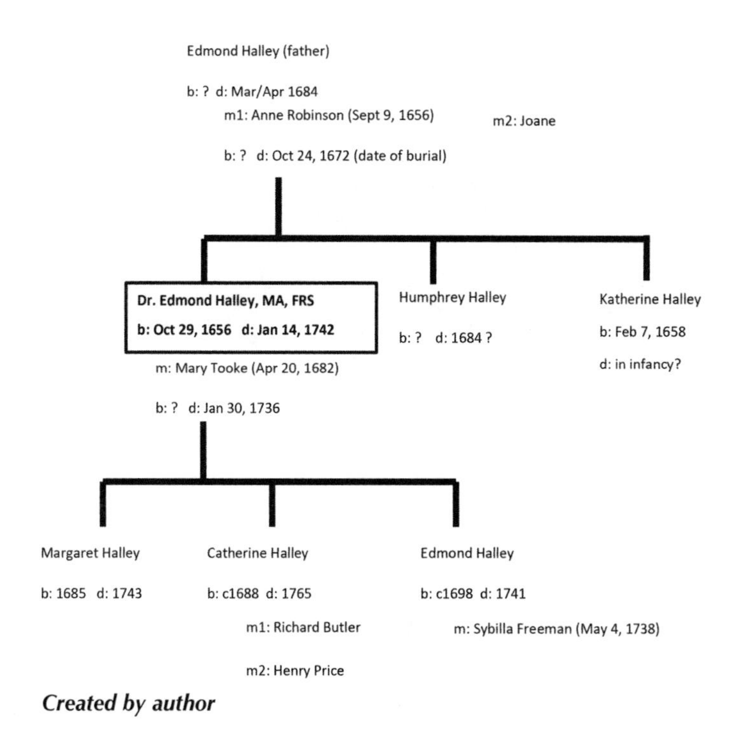

Created by author

Note: This information and the sources from which it came can be found in MacPike (1937, 1939). It can be seen that a number of dates are missing from the records.

APPENDIX B

Halley's Time Line

October 29, 1656 Edmond Halley is born.

November 1660 *The Royal Society is founded.*

1665 *The Great Plague of London takes place.*

1666 *The Great Fire of London occurs. St. Paul's School burns down.*

1670 or earlier Halley starts at St. Paul's School.

1671 St. Paul's School reopens in a new building on the original site next to the cathedral.

1672 Halley measures magnetic variation in London.

1672 Halley's mother dies.

Summer 1673 Halley starts at Oxford University (Queen's College).

1675 *King Charles II founds the Royal Observatory at Greenwich.*

1676 Halley writes a paper on the computation of planetary orbits.

November 1676 Halley sets off for St. Helena.

February 1677 Halley arrives at St. Helena.

October 1677 Halley observes the transit of Mercury across the Sun.

February 1678 Halley leaves St. Helena.

May 1678 Halley returns to London.

November 1678 Halley is made a Fellow of the Royal Society.

November 1678 Halley is awarded an MA by royal mandate from Charles II.

1679 Halley publishes a book on his St. Helena observations, *Catalogus Stellarum Australium.*

May-July 1679 Halley makes a diplomatic visit to the distinguished Polish astronomer Johannes Hevelius.

Late 1680-early 1682 Halley visits Paris, much of France, and Italy.

1682 Halley probably meets Isaac Newton.

April 1682 Halley marries Mary Tooke and moves to Islington (London).

1683 Halley publishes a paper on terrestrial magnetism.

January 1684 Halley, Wren, and Hooke discuss whether an inverse square law of gravitation could produce elliptical orbits.

March-April 1684 Halley's father is murdered. He has legal disputes with stepmother.

August 1684 Halley visits Newton in Cambridge to discuss the consequences of the inverse square law of gravitation.

1685 Charles II dies. His Catholic brother James II inherits the throne.

1685 The Edict of Nantes is revoked in France. As a result, the mathematician Abraham de Moivre moves to England shortly afterward and later becomes a friend of Newton and of Halley.

1685 Halley's first daughter, Margaret, is born.

1686-1698 Halley works for the Royal Society as its clerk.

1686 Halley publishes a paper entitled "Trade Winds and Their Causes."

1687 Halley publishes a paper entitled "Evaporation."

July 1687 Newton's *Principia* is published thanks entirely to Halley's efforts.

1688 Halley's second daughter, Catherine, is born.

1688 *The Glorious Revolution. King James II is overthrown, and replaced by William III of Orange and his wife Mary (daughter of James), who reigned jointly.*

1689 Christiaan Huygens arrives in London for a Royal Society meeting and meets Halley.

1691 Halley builds a diving bell.

1691 Halley publishes a paper on the date of Julius Caesar's arrival in Britain.

1691 Halley is refused the Savilian Chair of Astronomy at Oxford University.

1693 Halley becomes a founder of actuarial science when he calculates annuities in his Breslau table of mortality.

1694 Halley holds a presentation on the cause of the biblical flood. (Not published until 1724.)

1696–1698 Halley is deputy comptroller of the mint at Chester.

June 1697 Halley conducts an experiment with a barometer atop Mount Snowdon (close to Chester).

1698 A drunken Halley may or may not have pushed a drunken Peter the Great in a wheelbarrow around John Evelyn's garden and through his favorite hedge.

1698 Halley's son Edmond is born.

1698–1700 Halley is the captain in charge of the *Paramore* (two scientific expeditions to plot the Earth's magnetic field).

October 1698–July 1699 Halley makes his first voyage.

September 1699–September 1700 Halley makes his second voyage.

June 1701–October 1701 Halley surveys the tides in the English Channel.

1702 *King William III dies. Queen Anne inherits throne.*

1702–1703 Halley conducts surveys to advise on fortification of harbors of Trieste and Buccari (Bakar) in the Gulf of Venice.

1704 Halley is appointed Savilian Professor of Geometry at Oxford University.

1705 Halley publishes *A Synopsis of the Astronomy of Comets*, in which he makes the prediction that the 1682 comet will return in 1758.

1706 Halley publishes his translation of Apollonius's book *De Sectione Rationis*.

1707 *Act of Union between England and Scotland.*

1707 Four ships in the British fleet sink off the Isles of Scilly with huge loss of life, a disaster that Halley has warned about.

1710 Halley publishes his translation of Apollonius's book on conics.

1712 Halley prepares Flamsteed's observations of stellar positions for publication.

1713 Halley becomes secretary of the Royal Society.

1714 Halley publishes a paper proposing calculation of the age of the Earth from the salinity of the sea. He speculates that this would give an age well above the biblical 6,000 years.

1714 *Queen Anne dies and is succeeded by her second cousin and closest Protestant relative, George I of Hanover.*

April 22, 1715 Halley observes a total eclipse of the Sun in southern England.

1716 Halley publishes a paper suggesting how to use transit of Venus to calculate absolute distances in the solar system.

1718 Halley discovers that certain "fixed" stars have moved their positions relative to where they were in the time of the ancient Greeks.

1720 Halley is appointed as the second Astronomer Royal.

1721 Halley writes a paper suggesting that space must be infinite on the grounds that if it were not, gravitational collapse would inevitably happen.

1727 *Newton dies. George I dies and is succeeded by his son George II.*

1729 Queen Caroline visits the Greenwich Observatory.

1729 Halley is awarded Foreign Membership of the French Académie Royale des Sciences.

1731 Halley puts forward his proposal to find longitude at sea by means of lunar observations.

1734 Bishop George Berkeley's book *The Analyst* is first published. Its reference to the "infidel mathematician" is believed by many to have been a reference to Halley.

1735 Halley's wife Mary dies.

1738 or 1739 Halley suffers a minor stroke.

1740 Halley's son Edmond dies.

January 14, 1742 Halley dies at the age of 85. He is buried in the parish church of Lee.

1752 Britain finally catches up with the rest of Europe and changes from the Julian to the Gregorian calendar.

1758 Halley's Comet reappears, thereby guaranteeing him immortality.

1761 and 1769 A big international effort takes place to observe transits of Venus, as recommended by Halley.

2061 The next year in which Halley's Comet will appear.

APPENDIX C

Key Astronomical and Other Scientific Figures in and around the Time of Halley

ENGLAND

Edmond Halley (1656–1742)

Jonas Moore (1617–1679), patron of Halley

Christopher Wren (1632–1723), architect, astronomer, mathematician, and all-round genius

Joseph Williamson (1633–1701), patron of Halley

Robert Hooke (1635–1703), brilliant experimenter, overshadowed by Newton

Isaac Newton (1642–1727), author of *Principia*

John Flamsteed (1646–1719), first Astronomer Royal and a bitter enemy of Halley

Hans Sloane (1660–1753), president of the Royal Society after Newton from 1727 to 1741

Martin Folkes (1690–1754), president of the Royal Society from 1741 to 1752 and probable Halley biographer

SCOTLAND

James Gregory (1638–1675), brilliant mathematician who designed (but never built) a type of reflecting telescope
David Gregory (1659–1708), nephew of James Gregory who beat Halley to the post of Savilian Professor of Astronomy at Oxford University in 1691

FRANCE

René Descartes (1596–1650), philosopher who developed a vortex theory of planetary motion, among numerous other achievements
Abraham de Moivre (1667–1754), mathematician
Jean-Jacques d'Ortous de Mairan (1678–1771), Halley biographer

ITALY/FRANCE

Giovanni Domenico Cassini (1625–1712), head of the Paris Observatory

POLAND

Johannes Hevelius (1611–1687), leading observational astronomer

NETHERLANDS

Christiaan Huygens (1629–1695), leading European mathematician, physicist, and astronomer

APPENDIX D

Primary Source Documents

I have used several early sources to paint a picture of Halley's life and work, outlined here. *Memoir of Dr. Edmond Halley* and *Éloge de M. Halley* were published as part of *Correspondence and Papers of Edmond Halley*, but they are sufficiently important to justify separate mention.

Biographia Britannica, Volume 4. Edited by William Oldys (1696–1761) and others and published in seven folio volumes (1747–1766), this was a collection of lives of eminent people in Great Britain and Ireland from the earliest times to the mid-18th century, including, of course, Edmond Halley. At least some of the material on Halley was supplied by his son-in-law Henry Price. The book was digitized by Google.

Memoir of Dr. Edmond Halley. By an unknown author, probably Martin Folkes, who was president of the Royal Society from 1741 to 1752, this clearly was written shortly after Halley's death.

Éloge de M. Halley. By Jean-Jacques d'Ortous de Mairan, this is a brief biography written in French in which, fortunately, this author is reasonably fluent. The éloge was written shortly after Halley's death and read to the French Academy in 1742. It derives to some extent but by no means completely from *Memoir of Dr. Edmond Halley*, mentioned above.

Correspondence and Papers of Edmond Halley. By Eugene Fairfield MacPike (1932), this is an invaluable collection of early documents and letters by and about Halley.

Hevelius, Flamsteed, and Halley. This is about three contemporary astronomers and their mutual relations, by Eugene Fairfield MacPike (1937). Although not strictly a primary source itself, it is well researched and contains references to several primary sources.

Dr. Edmond Halley (1656–1742). This is a bibliographical guide to Halley's life and work arranged chronologically, by Eugene Fairfield MacPike (1939).

A Synopsis of the Astronomy of Comets. By Edmond Halley, the original version was written by Halley in Latin to ensure widespread circulation throughout Europe, but an English translation appeared in the same year (1705).

The History of the Royal Society of London, Volume 4. An invaluable source of information, this is the (downloadable) minutes of Royal Society meetings from 1680 to 1687. Unfortunately, after 1687, only hard copies are available.

An Account of the Rev'd John Flamsteed FRS. By Francis Bailey, originally published in 1835, this consists mainly of Flamsteed's own autobiographical notes as well as a significant number of his letters (and replies to these), many containing vicious criticisms of Halley.

The Correspondence of Isaac Newton, Volume II (1676–1687)

A Defence of Halley against the Charge of Religious Infidelity. By Rev. S. J. Rigaud, published 1844.

The Three Voyages of Edmond Halley in the Paramore. Edited by Norman J. Thrower and published 1981, this includes Halley's logbooks and numerous items of correspondence of the time.

A great number of scientific papers written by Halley and presented to the Royal Society are downloadable as PDF files directly from the Royal Society website. Relevant papers are referenced in the notes to each chapter of this book.

Although they are not primary source documents, there have been several papers about Halley presented to the Royal Society in the past few decades. These are also available on the Royal Society website (usually for a small fee). The ones I referred to are as follows (in alphabetical order by author):

T. E. Allibone, "Edmond Halley and the Clubs of the Royal Society" (1974)
C. Andrade, "Robert Hooke" (1950)
S. Chapman, "Edmond Halley, F.R.S." (1956)
Cohen and Ross, "The Commonplace Book of Edmond Halley" (1985)
Sir Alan Cook, "Edmond Halley and Newton's *Principia*" (1991)
Sir Alan Cook, "Halley the Londoner" (1993)
Charles H. Cotter, "Captain Edmond Halley, R.N., F.R.S." (1981)
Major Greenwood, "The First Life Table" (1938)
David Hughes, "The History of Halley's Comet" (1987)
David Hughes, "The *Principia* and Comets" (1988)
H. Spencer Jones, "Halley as an Astronomer" (1957)
Milo Keynes, "The Personality of Isaac Newton" (1995)
P. D. Lawrence and A. G. Molland, "David Gregory's Inaugural Lecture at Oxford" (1970)

Dmitri Levitin, "Halley and the Eternity of the World Revisited" (2013)

Bernard Lovell, "The RS, the RGO, and the Astronomer Royal" (1994)

Simon Schaffer, "Halley's Atheism and the End of the World" (1977)

D. T. Whiteside, "The Prehistory of the *Principia*" (1991)

D. T. Whiteside, "Newton's Marvelous Year" (1966)

APPENDIX E

Calculation of Inverse Square Law for a Circular Orbit (Assuming Only Kepler's Third Law—Only for Those Interested in Math!)

1. Christiaan Huygens had calculated that the inward acceleration, a, of a body in a circular orbit round the Sun was given by

$$a = v^2/r,$$

where v is its speed and r is the distance from the Sun. For a circular orbit, the distance d that the body travels in completing one orbit is the circumference of a circle of radius r,

$$d = 2\pi r,$$

and if it does this in time T (the period of the orbit), then

$$v = 2\pi r/T \text{ (i.e., speed = distance/time, by definition).}$$

Substituting for v, we get

$$a = 4\pi^2 r^2/r\,T^2. \text{ (i)}$$

2. Johannes Kepler had established in his third law of planetary motion that

$$r^3/T^2 = K, \text{ a constant,}$$

$$\text{or } r^3/K = T^2. \text{ (ii)}$$

3. Substituting (ii) into (i),

$$a = 4\,\pi^2\, r^2\, K/r\, r^3,$$

and combining all the constants into a single constant K' and canceling the extra values of r in the numerator, we get

$$a = K'/r^2. \text{ (iii)}$$

In other words, the acceleration experienced by a body in a circular orbit round the Sun is inversely proportional to the square of the distance from the Sun.

From Newton's third law, the force F on a body of mass m having an acceleration a is

$$F = m\,a.$$

So, by substituting for a from (iii),

$$F = m\,K'/r^2.$$

This shows that the gravitational force on a body in a circular orbit round the Sun is also inversely proportional to the square of the distance from the Sun.

Demonstrating this mathematically for an elliptical orbit (which is what Kepler had deduced from Tycho Brahe's observations were the orbits of the planets) is vastly more difficult. Halley and others failed, and only Newton succeeded.

APPENDIX F

Julian Dates versus Gregorian Dates

There was a problem with the old Julian calendar (which originated with Julius Caesar and his advisers), which had been abandoned by most of continental Europe in 1582 but was still in use in England even in the time of Halley.

The problem was a result of the fact that there aren't a whole number of days in a year, and any calendar based on the Sun needs to take account of that. The Julian calendar assumed that there are exactly 365¼ days in a year, but this isn't quite true. As a result, throughout the centuries, an increasing mismatch had arisen between the theoretical date of the summer solstice (June 21, the day when the Sun is meant to reach its highest point in the sky at midday) and its actual date (June 11 by 1582). If this had been allowed to continue, then after about 750 years, the summer solstice would eventually have fallen on December 21.

To resolve the problem, in 1582, Pope Gregory XIII ruled that October 5 that year would become October 15, correcting the cumulative error up to that time. To prevent further errors from arising in the future, he also ruled that the extra day that we insert in leap years (on February 29) would be omitted at the end of those centuries not divisible by 400.

This went a very considerable way to correcting the problem with the Julian calendar and was a thoroughly sensible reform.

England was well behind most of the rest of Europe, as it had still not adopted the Gregorian calendar when Halley was born and would not do so until some 10 years after Halley's death. (This was largely because the change was thought of as a wicked Catholic plot rather than the necessary reform of the calendar that it actually was.) So all dates used in this book, unless otherwise stated, are still those of the Julian calendar.

Occasionally, the date of Halley's birth is given in various sources (usually those that originated in mainland Europe) as November 8, 1656, because it has been converted to the Gregorian figure. Other dates in various sources sometimes differ by 10 or 11 days from the Julian date, again because they have also been stated in terms of the Gregorian calendar.

It should also be mentioned that, in Halley's time, for some purposes, a year was deemed to end on December 31 and for others on the following March 25. This meant that a date falling in that period, say, March 1, 1666, was usually expressed as March 1, 1665/6. I have not followed that convention in this book and have treated all years as beginning on January 1.

APPENDIX G

The Coriolis Effect

A full discussion of the Coriolis effect is outside the scope of this book; however, the case of air movements creating the trade winds can be understood in principle as follows.

Imagine two sisters playing catch while standing on a spinning roundabout. One child (Charlotte) is standing on the edge of the roundabout, and the other (Stella) is nearer the center. For Stella to pass the ball to Charlotte at the edge, she must aim ahead of where Charlotte is when she throws the ball, aiming instead at where Charlotte will be when the ball reaches the edge of the roundabout. (If she were to throw it directly at Charlotte, it would miss because of the roundabout's movement.) A third child (Oscar) standing in the playground will always see the ball move in a straight line, but to the children on the roundabout, it appears to describe a curved path, turning in the opposite direction to the roundabout's rotation. This apparent curvature is called the Coriolis effect.

Now imagine the roundabout enlarged and wrapped around the Earth's Northern Hemisphere, the center of the roundabout at the North Pole and the edge at the equator. The ball is replaced by a parcel of air and the children's throws by the effect of the convection Halley

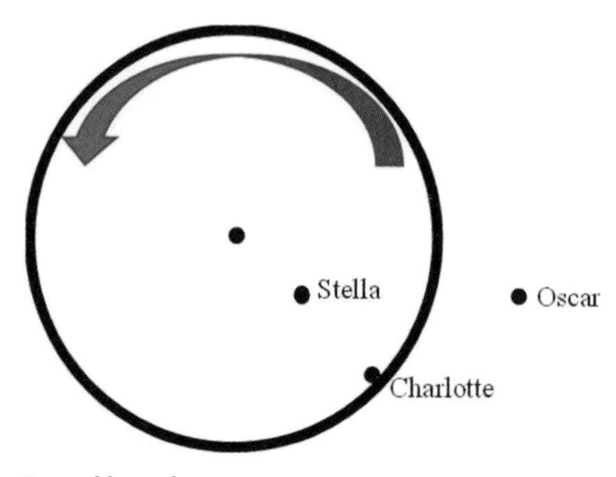

Created by author

described. A parcel of air at 30° N, say, over Algeria, is moving south toward Ghana. But Ghana (which is directly south of Algeria) is some 850 km farther away from the Earth's axis of rotation than Algeria because it is much nearer the equator. Like a ball thrown from Stella to Charlotte, the air will appear to follow a curved path out into the Atlantic west of Ghana. The effect is mitigated by friction with the ground, but enough remains that the southbound air parcel is now heading southwest. It has become a northeasterly trade wind.

It is terribly tempting to think that the deviation in the air's course is caused by the movement southward, but it isn't. It is caused by the movement outward from the axis. If we lived on a cylindrical Earth, there would be no trade winds.

ACKNOWLEDGMENTS

I am grateful to the following for their permission to use certain images:

The Royal Society: portrait of Edmond Halley
Euratlas maps: map of Europe in 1700
The Royal Society: cover page of Newton's *Principia*
The Royal Society: Halley's map of the trade winds
Science Museum: Halley's Diving Bell
David Higham Associates: two maps of Halley's voyages around the Atlantic
Scilly Isles Museum and Sandra Greaves: two items in the museum
Institute of Astronomy Library, University of Cambridge: Halley's eclipse map
Royal Greenwich Observatory: location of Halley's tombstone in the grounds of the RGO

I have also used images from the following:

European Southern Observatory (ESO): for Omega Centauri
NASA's NSSDC Photo Gallery: for Halley's Comet

NASA, ESA, S. Beckwith (STScI), and the Hubble Heritage Team (STScI/AURA): for M51

New York Public Library Digital Collections. Accessed November 7, 2022. https://digitalcollections.nypl.org/items/510d47da-ef24-a3d9-e040-e00a18064a99: for Halley's chart of magnetic variations over the Atlantic Ocean.

All other photographs and diagrams were taken or produced by me or my daughter Rachel or taken from Wikimedia Commons.

To the best of my knowledge, I have obtained all necessary permissions for the contents of this book. If there are any errors or omissions, Rowman & Littlefield will be pleased to insert the appropriate acknowledgment in any subsequent printing of this book.

A NOTE ABOUT THE ENDNOTES

A complete list of the primary source documents used for this book is set out in appendix D. Those that have been used most frequently have, for convenience, been abbreviated as follows in the immediately following chapter end notes:

BB Vol. 4: *Biographia Britannica*, Volume 4. (Digitized by Google)

Memoir: *Memoir of Dr. Edmond Halley*, probably by Martin Folkes, written very shortly after Halley's death, and included in *Correspondence and Papers of Edmond Halley*, by Eugene Fairfield MacPike, published 1932.

Éloge: *Éloge de M. Halley*, by Jean-Jacques d'Ortous de Mairan, also written very shortly after Halley's death, and included in *Correspondence and Papers of Edmond Halley*, by Eugene Fairfield MacPike, published 1932.

MacPike (1932): *Correspondence and Papers of Edmond Halley*, by Eugene Fairfield MacPike, published 1932.

MacPike (1937): *Hevelius, Flamsteed, and Halley*, by Eugene Fairfield MacPike, published 1937.

MacPike (1939): *Dr. Edmond Halley: A Biographical Guide to His Life and Work*, by Eugene Fairfield MacPike, published 1939.

Synopsis: *A Synopsis of the Astronomy of Comets*, by Edmond Halley.

RS Vol. 4: *The History of the Royal Society of London, 1680–1687*, Volume 4.

Flamsteed: *An Account of the Rev'd John Flamsteed FRS*, by Francis Bailey.

Rigaud: *A Defence of Halley against the Charge of Religious Infidelity*, by the Rev'd S. J. Rigaud.

NOTES

CHAPTER 1

1. BB Vol. 4, 2494.
2. MacPike (1939, 25).
3. MacPike (1939, 26).
4. We simply don't know exactly when. Details of his early life are infuriatingly lacking. There is a story that an apprentice of his father taught him writing and arithmetic before he arrived at St. Paul's. See MacPike (1939, 26).
5. I am grateful to Ginny Dawe-Woodings, the St. Paul's School archivist, for providing what little background information there is on Halley's time at St. Paul's.
6. MacPike (1932, 1).
7. BB Vol. 4, 2494.
8. "The Commonplace Book of Edmond Halley," 3.
9. The original paper was published by the Royal Society in its *Philosophical Transactions* for October 1676.
10. MacPike (1932, 37).

11. The Royal Museums Greenwich website gives the story at https://www.rmg.co.uk/stories/blog/curatorial/founding-royal-observatory.

12. A detailed explanation of the lunar distance method is given at https://en.wikipedia.org/wiki/Lunar_distance_(navigation).

13. Flamsteed, 111–12.

14. The East India Company was essentially a device for transferring wealth from India to England, by force if necessary. See William Dalrymple's book *The Anarchy*.

15. MacPike (1932, 40).

16. MacPike (1937, 40).

17. BB Vol. 4, 2495. In addition to the slightly weaker gravity due to being closer to the equator, there was also a further slight weakening due to Halley's clock being an estimated 800 meters above sea level. See RS Vol. 4, 189.

18. *The History of the Royal Society of London, Volume 3*, 409.

19. A brief summary of various scientific societies is at https://thonyc.wordpress.com/2010/02/18/it-wasnt-the-first-but.

20. MacPike (1932, 43).

21. "Hospitem gratissimum, Virum integerrimum et veritatis amantissimum." MacPike (1932, 4).

22. MacPike (1937, 113).

23. See Winterburn (2003, 11), Cook (1998, 326), and Lancaster-Brown (1985, 32).

24. BB Vol. 4, 2495, note l.

25. MacPike (1932, 48).

26. MacPike (1932, 49).

27. MacPike (1932, 48).

28. MacPike (1937, 116).

29. MacPike (1932, 52).

30. Memoir, 5.

31. Éloge, 19, author's translation.

CHAPTER 2

1. MacPike (1937, 50). The words are those of Augustus de Morgan (1806–1871).

2. BB Vol. 4, 2504. Letter from Halley to Newton dated June 29, 1686:

> And this I know to be true, that in January 1684, I having from the consideration of the sesquialterate proportion of Kepler, concluded that the centripetal force decreased in the proportion of the squares of the distances reciprocally, came one Wednesday to town (from Islington) where I met with Sir Christopher Wren and Mr Hooke, and falling in discourse about it, Mr Hooke affirmed, that upon that principle all the laws of the celestial motions were to be demonstrated and that he himself had done it. I declared the ill success of my attempts, and Sir Christopher to encourage the enquiry said, that he would give Mr Hooke or me two months' time to bring him a convincing demonstration thereof, and besides the honour, he of us that did it, should have from him a present of a book of forty shillings.

3. RS Vol. 4, 244–52.

4. Westfall (1980, 393).

5. MacPike (1932, 177–78). In contradiction, Alan Cook's sources indicate that the whole of the £100 was paid to the boy who found the body, £20 directly and £80 in trust.

6. Joane remarried on June 2, 1685, a little under 14 months after the death of Halley's father was discovered. See "Edmond Halley and Newton's Principia," 131, by Alan Hugh Cook, a Royal Society paper published July 1, 1991.

7. This is a famous quotation that appears, for example, in the 1988 Royal Society paper by David Hughes titled "Principia and Comets." The original document containing the quote is held in the Joseph Halle Schaffner Collection, University of Chicago Library.

8. *The History of the Royal Society of London, Volume 3*, 1.

9. Hooke's criticisms can be found in the Royal Society minutes, Volume 3, 10ff. (February 1672).

10. There are some brief references by Hooke to this correspondence in the Royal Society minutes, Volume 3, 519 (December 1679), and Volume 4, 1 (January 1680).

11. See this episode of BBC Radio 4's *In Our Time* on Robert Hooke at https://www.bbc.co.uk/programmes/b070h6ww.

12. RS Vol. 4, 347.

13. RS Vol. 4, 454.

14. RS Vol. 4, 453.

15. RS Vol. 4, 455.

16. See "Edmond Halley and Newton's Principia," 134.

17. RS Vol. 4, 545:

> The question being put, whether Mr. Halley should have fifty copies of the History of Fishes instead of the fifty pounds ordered him by the last meeting of the council, comprehending the twenty books formerly put into the hands of Mr. Smith the bookseller, it was determined by ballot in the affirmative.
>
> It was ordered, that Mr Halley receive a gratuity of twenty other copies of the History of Fishes, in consideration of his arrears in the last year ending January 27, 1687.

18. RS Vol. 4, 489.

19. RS Vol. 4, 505.

20. RS Vol. 4, 516.

21. RS Vol. 4, 479–80.

22. RS Vol. 4, 484.

23. RS Vol. 4, 484.

24. Extract from Isaac Newton's letter dated June 20, 1686:

> For tis plain by his words he knew not how to go about it. Now is not this very fine? Mathematicians that find out, settle and do all the business must content themselves with being nothing but dry calculators & drudges & another that does nothing but pretend & grasp at all things must carry away all the invention as well of those that were to follow him as of those that went before.

25. Also from Isaac Newton's letter dated June 20, 1686: "The third [book] I now design to suppress."

26. See in particular Halley's reply to Newton of June 29, part of which is given in note 2 above.

27. RS Vol. 4, 486.

28. MacPike (1932, 81).

29. See Halley's letter of July 5 to Newton.

30. BB Vol. 4, 2505. There is also a letter from Halley to James II dated July 1687 and reprinted in MacPike (1932, 85).

31. MacPike (1937, 50).

32. Quote from p. 4 of the edition of *Principia* published by Prometheus Books in 1995.

33. MacPike (1932, 132).

CHAPTER 3

1. White (1993, 1:174–83).

2. Murdin and Penston (2004, 83).

3. BBC talk by Lisa Jardine, https://www.bbc.co.uk/news/magazine-21802843.

4. White (1993, 1:181).

5. Love (2015, 130–31).

6. RS Vol. 4, 62, 67.

7. BB Vol. 4, 2499.

8. *Principia*, 401–2. "Cor 2: But their orbits will be so near to parabolas that parabolas may be used for them without sensible error."

9. MacPike (1932, 92).

10. MacPike (1932, 92).

11. MacPike (1932, 92).

12. Halley (1705, 22).

13. Lancaster-Brown (1985, 87).

14. Murdin and Penston (2004, 178).

15. H. U. Keller, "Comets—Dirty Snowballs or Icy Dirtballs?," in Hunt and Guyeme (1989, 39–45).

16. Murdin and Penston (2004, 97).

CHAPTER 4

1. Halley's paper on the trade winds can be found at https://royalsocietypublishing.org/doi/10.1098/rstl.1686.0026.

2. A description of Hadley's (not Halley's) correct explanation can be found at https://rmets.onlinelibrary.wiley.com/doi/epdf/10.1002/wea.228.

3. Halley's paper on evaporation was titled "An Estimate of the Quantity of Vapour Raised Out of the Sea by the Warmth of the Sun; Derived from an Experiment Shown before the Royal Society, at One of Their Late Meetings" and can be found at https://doi.org/10.1098/rstl.1686.0067.

4. Royal Society minutes are, unfortunately, not available on the internet for dates later than 1687. This is a transcript of the hard-copy minutes of August 26, 1691, as they appear in the Society's Journal Book and as

quoted in MacPike (1932, 224). The detailed papers themselves can be found in MacPike (1932, 150–56).

5. Also as quoted in MacPike (1932, 224).

6. https://en.wikipedia.org/wiki/History_of_underwater_diving.

7. Halley's paper was titled "An Estimate of the Degrees of the Mortality of Mankind; Drawn from Curious Tables of the Births and Funerals at the City of Breslau; With an Attempt to Ascertain the Price of Annuities upon Lives" and can be found at https://doi.org/10.1098/rstl.1693.0007.

8. Information on John Graunt can be found at https://www.theactuary.com/archive/old-articles/part-3/2012/09/21/who-was-captain-john-graunt; see also Major Greenwood, "The First Life Table," https://royalsocietypublishing.org/doi/10.1098/rsnr.1938.0017.

9. David R. Bellhouse, "A New Look at Halley's Life Table," *Journal of the Royal Statistical Society, Series A (Statistics in Society)* 174, no. 3 (2011): 823–32, http://www.jstor.org/stable/23013523 (accessed April 28, 2022).

10. BB Vol. 4, 2512, note HH. See also MacPike (1937, 55).

11. Halley's paper on the flood can be found at https://royalsocietypublishing.org/doi/10.1098/rstl.1724.0023.

12. "A Letter from Mr. Halley of June the 7th. 97. concerning the Torricellian Experiment Tryed on the Top of Snowdon-Hill and the Success of It," https://doi.org/10.1098/rstl.1695.0106.

13. Halley's paper of 1686 on the use of the barometer can be found at https://royalsocietypublishing.org/doi/10.1098/rstl.1686.0017.

14. RS Vol. 4, 491.

15. MacPike (1932, 103).

16. MacPike (1932, 102).

CHAPTER 5

1. BB, Vol. 4, 2517.
2. Éloge, 27, author's translation.
3. The scientific journal *Nature* 21 (1880): 303.
4. MacPike (1932, 5).
5. MacPike (1932, 76–77).
6. Flamsteed, 213.
7. Flamsteed, 293.
8. Flamsteed, 150.
9. Flamsteed, 194.
10. Flamsteed, 244.
11. Flamsteed, 289.
12. A fuller quotation is "and that I have no esteem of a man who has lost his reputation, both for skill, candor and ingenuity, by silly tricks, ingratitude, and foolish prate: and that I value not at all, or any of the shame of him and his infidel companions; being very well satisfied that if Christ and his apostles were to walk again upon earth, they should not escape free from the calumnies of their venomous tongues. But I hate his ill manners, not the man: were he either honest, or but civil, there is none in whose company I could rather desire to be" (Flamsteed, 133).
13. The two most relevant papers on the subject that were published by the Royal Society were those of Simon Schaffer ("Halley's Atheism and the End of the World") in 1977 and Dmitri Levitin ("Halley and the Eternity of the World Revisited") in 2013. They can be found on the Royal Society website at https://royalsocietypublishing.org/doi/10.1098/rsnr.1977.0004 and https://royalsocietypublishing.org/doi/10.1098/rsnr.2013.0019, respectively.
14. MacPike (1932, 264).
15. MacPike (1937, 73).

16. By Bishop Rawlinson, according to MacPike (1937, 72). See also MacPike (1932, 264–65), note 10.

17. See the Royal Society paper "David Gregory's Inaugural Lecture at Oxford" by P. D. Lawrence and A. G. Molland, published in 1970.

18. MacPike (1932, 265).

19. For example, Rigaud (1844, 24ff.).

20. See https://en.wikipedia.org/wiki/Savilian_Professor_of_Astronomy.

21. In a letter to Abraham Hill dated June 22, 1691, Halley asks Hill to intercede with Tillotson to postpone the election for astronomy professor "till I can shew that I am not guilty of asserting the eternity of the world" (MacPike 1932, 88).

22. MacPike (1932, 264). From table talk of Bishop Hough, who was a later successor to Bishop Stillingfleet as Bishop of Worcester. It is not implausible that this story could have been passed down from bishop to bishop. A similar account elsewhere (also quoted on p. 264) in which Halley is reported to have said that "he believed in God and that was all" adds weight to the story if the latter is indeed an independent source.

23. Rigaud (1844, 8), quoting what Whiston had recorded.

24. MacPike (1932, 221).

25. A recent survey (2018) by Rice University and others showed that scientists in the United Kingdom are significantly more likely (45 percent vs. 18 percent) to be atheists than the general population. See https://www.sciencedaily.com/releases/2018/12/181219115525.htm.

CHAPTER 6

1. S. P. Oliver, "Proposed Monument to Halley," *The Observatory* 35 (1880):349. For the justification for this comment, see Thrower (1981, 69ff.).

2. Arthur MacGregor, "The Tsar in England: Peter the Great's Visit to London in 1698," *The Seventeenth Century* 19, no. 1 (2004): 127, doi:10.1080/0268117X.2004.10555538.

3. Cross (2001, 22–25).

4. Flamsteed, 65.

5. BB Vol. 4, 2517.

6. Memoir, 7–8.

7. Éloge, 26.

8. Cross (2001, 18). See also Leo Loewenson, "People Peter the Great Met in England: Moses Stringer, Chymist and Physician," *The Slavonic and East European Review* 37, no. 89 (June 1959): 459–68 (p. 5 of the account).

9. At the time of the 1910 reappearance of Halley's Comet, the story appeared (with no indication of its source) in the March 12, 1910, edition of *The Spectator*, a right-wing English weekly magazine, as part of an account of Halley's life and his comet. The story is repeated by MacPike (1937, 57), but its doubtful authority is mentioned.

10. Thrower (1981, 262). I am grateful to David Higham Associates for giving permission to quote this letter.

11. MacPike (1932, 244).

12. MacPike (1932, 29).

13. BB Vol. 4, 2494, note A.

14. MacPike (1932, 109).

15. MacPike (1932, 113).

16. Royal Society Journal Book, Vol. VIII, 227.

17. MacPike (1932, 114).

18. Thrower (1981, 49).

19. BB Vol. 4, 2502, note O. See also MacPike (1932, 246–47).

20. Thrower (1981, 321).

21. MacPike (1932, 249).

CHAPTER 7

1. This is a famous quote. It appears on page 215 in one of the letters of Flamsteed in Baily (1835).

2. BB Vol. 4, 2513, section KK.

3. A letter of 1707 from de Moivre to Johann Bernoulli reads, "M. Halley vous fait ses complimens; il travaille à nous donner le 5e, 6e, 7e livre d'Apollonius "de conicis" qu'il a traduits de l'arabe sur une copie d'un excellent manuscrit qui est entre les mains du Primat d'Irlande. Il doit avoir aussi ce manuscrit pour le confronter avec la copie; il espère de rétablir le 8e livre, parce que le 7e contient tous les lemmes des propositions du 8e. Ce que j'en ai vu me paroit d'une beauté et d'une simplicité très grande."

4. BB Vol. 4, 2513, section LL.

5. Much of this information is in Larn (2006), published by the Council of the Isles of Scilly in 2007 to mark the 300th anniversary of the disaster.

6. Personal communication from Richard Larn.

7. "An advertisement [in the archaic sense of a warning] necessary for all navigators bound up the Channel of England," https://royalsocietypublishing.org/doi/10.1098/rstl.1700.0059. The article doesn't have his name appended, but his fingerprints are all over it.

8. MacPike (1932, 212).

9. Halley's paper on the saltiness of the oceans is available at https://doi.org/10.1098/rstl.1714.0031.

10. Murdin and Penston (2004, 133).

11. Jay Pasachoff, "Halley & His Maps of the Total Eclipses of 1715 and 1724," *Astronomy & Geophysics* 40, no. 2 (April 1999): 218–21.

12. Fred Espenak produces excellent maps forecasting the paths of future total solar eclipses at https://eclipse.gsfc.nasa.gov.

13. Halley's paper on the eclipse is available at https://royalsociety publishing.org/doi/10.1098/rstl.1714.0025.

14. "By these observations I am satisfied the Moon's latitude was less, by a minute, than I made it in the figure I sent you; and that the total eclipse reached near to Deal or Dover: whereas Dr Halley makes all East Kent free of them." Letter to Abraham Sharp, April 23, 1715, in one of the letters of Flamsteed in Baily (1835, 313).

15. A good account of how the distance of Venus can be established using this method can be found at https://www.youtube.com/watch ?v=GwP8wCzbFLc.

16. Halley's paper on the movement of certain "fixed" stars is available at https://doi.org/10.1098/rstl.1717.0025.

17. John C. Brandt, Department of Physics and Astronomy, University of New Mexico, https://ui.adsabs.harvard.edu/abs/2010JAHH...13 ..149B/abstract.

18. "Why Halley Did Not Discover Proper Motion and Why Cassini Did," https://journals.sagepub.com/doi/full/10.1177/00218286 19877967.

CHAPTER 8

1. Halley's first paper on what subsequently became known as Olbers' paradox is available at https://royalsocietypublishing.org/doi /epdf/10.1098/rstl.1720.0006; Halley's second paper is available at https://royalsocietypublishing.org/doi/epdf/10.1098/rstl.1720.0007.

2. Love (2015, 141–43).

3. Harrison (1987).

4. BB Vol. 4, 2515–16.

5. Halley's paper on finding longitude at sea is available at https:// royalsocietypublishing.org/doi/10.1098/rstl.1731.0031.

6. The story, albeit biased against the lunar distance method, is told in Sobel (1995).

7. Flamsteed refers to a visit by Halley to Child's Coffee House in a letter of November 1712 in Flamsteed, 297.

8. The RS paper on the Royal Society Club is available at https://royalsocietypublishing.org/doi/10.1098/rsnr.1971.0007; see also MacPike (1932, 252–54).

9. BB Vol. 4, 2516.

10. MacPike (1932, 26–27, author's translation).

CHAPTER 9

1. See Webb (1999, 35–45).

BIBLIOGRAPHY

Alter, Dinsmore. *Pictorial Guide to the Moon*. London: Arthur Barker, 1964.

Baily, Francis. *An Account of the Revd. John Flamsteed FRS*. London: Printed by order of the Lords Commissioners of the Admiralty, 1835.

Berry, Arthur. *A Short History of Astronomy*. New York: Dover, 1961.

Bronowski, Jacob. *The Ascent of Man*. London: BBC, 1976.

Carr, Bernard, ed. *Universe or Multiverse?* Cambridge: Cambridge University Press, 2007.

Cook, Alan. *Edmond Halley: Charting the Heavens and the Seas*. Oxford: Clarendon Press, 1998.

Cross, Anthony. *Peter the Great through British Eyes: Perceptions and Representations of the Tsar since 1698*. Cambridge: Cambridge University Press, 2001.

Dalrymple, William. *The Anarchy*. London: Bloomsbury, 2019.

Ehrman, Bart. *The New Testament*. Oxford: Oxford University Press, 2008.

Ellis, Richard S. *When Galaxies Were Born*. Princeton, NJ: Princeton University Press, 2022.

Fried, Michael N. *Edmond Halley's Reconstruction of the Lost Book of Apollonius's Conics*. New York: Springer, 2012.

Greene, Brian. *The Elegant Universe*. London: Jonathan Cape, 1999.

Guth, Alan H. *The Inflationary Universe*. London: Vintage, 1998.

Halley, Edmond. *A Synopsis of the Astronomy of Comets*. London: John Senex, 1705.

Harari, Yuval Noah. *Sapiens*. London: Vintage, 2011.

Harrison, Edward. *Darkness at Night*. Cambridge, MA: Harvard University Press, 1987.

Hoskin, Michael, ed. *The Cambridge Concise History of Astronomy*. Cambridge: Cambridge University Press, 1999.

Hunt, J., and D. Guyeme, eds. *Proceedings of an International Workshop on Physics and Mechanics of Cometary Materials*. ESA SP-30. Paris: European Space Agency, 1989.

Jennings, Charles. *Greenwich*. London: Abacus, 1999.

Jones, Sir Harold Spencer. *The Royal Observatory, Greenwich*. London: Longmans, 1943.

Koestler, Arthur. *The Sleepwalkers*. London: Penguin, 1964.

Krauss, Lawrence M. *A Universe from Nothing*. London: Simon & Schuster, 2012.

Lancaster-Brown, Peter. *Halley and His Comet*. Poole: Blandford Press, 1985.

Larn, Richard, ed. *Poor England Has Lost So Many Men*. St. Mary's: Council of the Isles of Scilly, 2006.

Love, David K. *Kepler and the Universe*. Amherst, NY: Prometheus Books, 2015.

MacPike, Eugene Fairfield. *Correspondence and Papers of Edmond Halley*. Oxford: Clarendon Press, 1932.

——. *Hevelius, Flamsteed, and Halley*. London: Taylor & Francis, 1937.

——. *Bibliographical Guide to Edmond Halley*. London: Taylor & Francis, 1939.

McBride, Neil, and Iain Gilmour, eds. *An Introduction to the Solar System*. Cambridge: Cambridge University Press, 2004.

McBride, Peter, and Larn, Richard. *Admiral Shovell's Treasure and Shipwreck in the Isles of Scilly*. London: Shipwreck & Marine, 1999.

Mitton, Simon, ed. *The Cambridge Encyclopedia of Astronomy*. London: Jonathan Cape, 1977.

Moore, Patrick. *The Solar System*. London: Methuen, 1958.

Murdin, Paul, and Penston, Margaret, eds. *The Canopus Encyclopedia of Astronomy*. Bristol: Canopus, 2004.

Nature (scientific journal) 21 (1880): 303.

Newton, Isaac. *Principia*. Amherst, NY: Prometheus Books, 1995.

Rees, Martin J. *Before the Beginning: Our Universe and Others*. London: Simon & Schuster, 1997.

Ridpath, Ian, ed. *Norton's 2000.0 Star Atlas*. New York: Longmans, 1989.

Rigaud, Rev. S. J. *A Defence of Halley against the Charge of Religious Infidelity*. Oxford: Ashmolean Society, 1844.

Ronan, Colin A. *Edmond Halley: Genius in Eclipse*. London: Macdonald, 1970.

Russell, Bertrand. *History of Western Philosophy*. London: Unwin, 1969.

Smart, William M. *Spherical Astronomy*. Cambridge: Cambridge University Press, 1965.

Sobel, Dava. *Longitude*. London: Fourth Estate, 1995.

Tegmark, Max. *Our Mathematical Universe*. London: Allen Lane, 2014.

Thiel, Rudolph. *And There Was Light: The Discovery of the Universe*. London: André Deutsch, 1958.

Thrower, Norman J., ed. *The Three Voyages of Edmond Halley in the Paramore, 1698–1701*. London: Hakluyt Society, 1981.

Vilenkin, Alex. *Many Worlds in One: The Search for Other Universes.* New York: Hill and Wang, 2006.

Webb, Stephen. *Measuring the Universe.* Chichester: Springer, 1999.

Westfall, Richard S. *Never at Rest: A Biography of Isaac Newton.* Cambridge: Cambridge University Press, 1980.

——. *The Life of Isaac Newton.* Cambridge: Cambridge University Press, 1993.

White, Andrew. *History of the Warfare of Science with Theology in Christendom.* 2 vols. Amherst, NY: Prometheus Books, 1993.

Winterburn, Emily. *The Astronomers Royal.* London: National Maritime Museum, 2003.

INDEX

Page references for figures are italicized.